THIS BOOK
IS A PLANT

THIS
BOOK
IS A
PLANT

How to Grow, Learn and
Radically Engage
with the Natural World

PROFILE BOOKS

wellcome
collection

This paperback edition first published in 2023

First published in Great Britain in 2022 by
Profile Books Ltd
29 Cloth Fair
London
ECIA 7JQ
www.profilebooks.com

Published in association with Wellcome Collection

183 Euston Road
London NWI 2BE
www.wellcomecollection.org

Text design by Crow Books

1 3 5 7 9 10 8 6 4 2

Printed and bound in Great Britain by
CPI Group (UK) Ltd, Croydon CRo 4YY

A CIP catalogue record for this book is available from the British Library.

ISBN 978 1 78816 692 8
eISBN 978 1 78283 799 2

CONTENTS

NOTE ON ILLUSTRATIONS

The chapter-opening illustrations, *Photosynthetics* (*Foto-sintéticos*), are a new body of work by the Argentinian artist Eduardo Navarro. These expansive and contemplative drawings of part-human, part-plant beings were commissioned by Wellcome Collection and La Casa Encendida, in partnership with Delfina Foundation, for exhibition in London and Madrid.

Navarro describes his work as emotional technology, a tool that allows us to develop trust, empathy and contemplation with non-human entities. The drawings were made with charcoal on handmade, biodegradable paper envelopes. Inside, seven seeds of an Ombu tree remain dormant. After being exhibited, the drawings will be placed in an open landscape, allowing nature to take over and activate the seeds within. While the seeds wait for the right humidity and light to begin their journey in time, the drawings act as a degradable womb, hosting their uncertain future.

THIS BOOK
IS A PLANT

VEGETAL TRANSMUTATION

Eduardo Navarro & Michael Marder

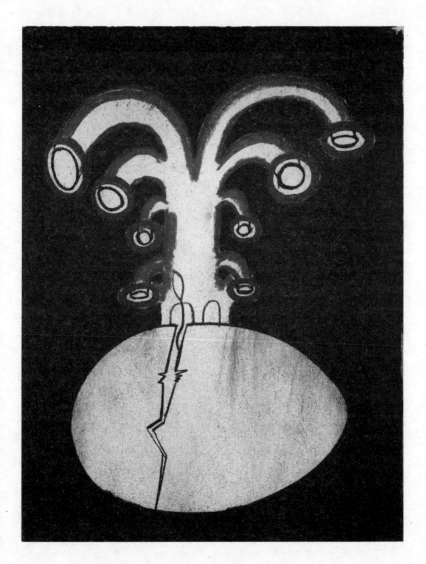

Please use these performative instructions to explore the world around you as you hold this book in your hands.

TAKE A MOMENT TO DRAW a cosmic breath with your whole body, slower than any breath you have ever taken in your life.

Close your eyes. See and hear with your skin as you embody the density that emanates from within the seed of your thoughts. Register the vibrations rippling throughout the space around you and the radiant waves enveloping you. Caress the air with your hands, petting its flows with each fingertip. Feel how grounded, yet free, you are. Stable, but supple.

Your skin is more than skin. It is at the same time an enormous leaf, in which your whole body is wrapped, as well as a respiratory system and a number of sense organs, photosensitive and acoustically engaged.

Break out into light, while staying connected to the dark and the obscure. Your fingers are roots and branches. There are more fingers branching out from the fingers in order for you to hug the earth and the sky better, more thoroughly.

One arm stretches tall; the other seeks deep in the soil. Span these extremes with your midsection: your chest, your trunk. Pay equal attention to both arms, orientations, worlds. Receive the kisses of the sun on your eyelids, ears, mouth, cheeks, fingers ... Imbibe the cool moistness of the earth with the other arm and a part of your trunk. Become the crossing between the warmth that gently touches you and the coolness you soak up.

Your limbs rotate in every direction, weightlessly reaching out to light and to darkness. Grow little by little, both intensively in time and extensively in space. Contract, minimise yourself, decay. Grow from contracting; contract from growing.

Focus on the soles of your feet as they absorb energy from planet Earth with every breath. Concentrate on the crown of your head, receiving the energy of the atmosphere. More like a tree crown, it is no longer a head. Think as you are breathing and perceiving – with your whole body, skin, limbs, lips, tips and edges. Do not hoard thoughts in your brain. Instead, let them circulate, like sap, in every part of yourself.

You are in an endless state of communion and infinite contemplation with other natural elements and beings. Can you see with your skin and hear with your arms? Can you think together with the air and the sun and the soil? Can you dream with your feet? Imagine with your fingertips?

Perhaps there are other plants sharing the space with you. Acknowledge these vegetal beings as you gradually move around in a wordless communicative practice. Experiment with the language of gestures, of physical expressions and forms. A choreography of touching without touching: the language of plants.

Move in the place where you are and sense the place moving along with you, growing and contracting rhythmically. Care for your place and for your attachment to it. You do not have your place; the place embraces you in itself. Render it equally welcoming, open to others. Experience your place as always the same *and* radically different, ramified and attuned otherwise in each instant. What does it mean for you to be at home there?

Your breathing is in a constant dialogue with your surroundings. Breathe in your depths, with the lungs, and on the surface, with the skin. Re-establish the ties between inner and outer breathing to sustain a respiratory conversation with the world. Root in your body through this double breath.

Inhale the space,
exhale the sky.

You are tending to your impulses, decisions, and actions on a horizontal plane, letting them become a holistic garden. Nurture them with your breath and the humid density of the obscure ground, from which they spring.

As you are very slowly dying, while also staying alive, your body becomes the soil you are living in. You are perpetually feeding yourself and others in a delicate equilibrium.

Feel the slow rotations of the cycle of life. In tandem with it, let fresh sprouts, bulbs and seeds in the space gradually take over. Nourish and support them from below. Allow the subtle movements of air and visible or invisible winged creatures to carry your seeds and pollen like messages across time. Disseminate yourself, holding nothing back.

Repeat once the book is closed,
weightlessly reaching out to light and to
darkness.

2

BEFORE ROOTS

Merlin Sheldrake

SOME TIME AROUND 600 million years ago, green algae began to move out of shallow fresh waters and onto the land. These were the ancestors of all land plants. The evolution of plants transformed the planet and its atmosphere and was one of the pivotal transitions in the history of life – a profound breakthrough in biological possibility. Today, plants make up 80 per cent of the mass of all life on Earth and are the base of the food chains that support nearly all terrestrial organisms.

Before plants, land was scorched and desolate. Conditions were extreme. Temperatures fluctuated wildly and landscapes were rocky and dusty. There was nothing that we would recognise as soil. Nutrients were locked up in solid rocks and minerals, and the climate was dry. This isn't to say that land was completely devoid of life. Crusts made up of photosynthetic bacteria, extremophile algae and fungi were able to make a living in the open air. But the harsh conditions meant that life on Earth was overwhelmingly an aquatic event. Warm, shallow seas and lagoons teemed with algae and animals. Sea scorpions several metres long ranged the ocean floor. Trilobites ploughed silty seabeds using spade-like snouts. Solitary corals started to form reefs. Molluscs thrived.

Despite its comparatively inhospitable conditions, land provided considerable opportunities for any photosynthetic organisms that could cope. Light was unfiltered by water, and carbon dioxide was more accessible – no small incentives for organisms that make a living by eating light and carbon dioxide. But the algal ancestors of land plants had no roots, no way to store or transport water, and no experience in extracting nutrients from solid ground. How did they manage the fraught passage onto dry land?

When it comes to piecing together origin stories it's difficult to find agreement among scholars. Evidence is usually sparse, and what fragments there are can often be mobilised to support different points of view. And yet, amid the slow-burning disputes that surround the early history of life, one piece of academic consensus stands out: it was only by striking up new relationships with fungi that algae were able to make it onto land.

These early alliances evolved into what we now call mycorrhizal relationships. Today, more than 90 per cent of all plant species depend on mycorrhizal fungi. Mycorrhizal associations are the rule, not the exception: a more fundamental part of planthood than fruit, flowers, leaves, wood or even roots. Out of this intimate partnership – complete with co-operation, conflict and competition – plants and mycorrhizal fungi enact a collective flourishing that underpins our past, present and future.

For the relationship to thrive, both plant and fungus must make a good metabolic match. In photosynthesis, plants harvest carbon from the atmosphere and forge the energy-rich carbon compounds – sugars and lipids – on which much of the rest of life depends. By growing within plant roots, mycorrhizal fungi acquire privileged access to these sources of energy: they get fed. However, photosynthesis isn't enough to support life. Plants and fungi need more than a source of energy. Water and minerals must be scavenged from the ground – full of textures and micropores, electrically charged cavities and labyrinthine rot-scapes. Fungi are deft rangers in this wilderness and can forage in a way that plants can't. By hosting fungi within their roots, plants gain hugely improved access to these sources of nutrients. They, too, get fed. By partnering, plants gain a prosthetic fungus, and fungi gain a prosthetic plant. Both use the other to extend their reach.

It isn't clear how mycorrhizal relationships first arose. Some venture that the earliest encounters were soggy, disorganised affairs: fungi seeking food and refuge within algae that washed up onto the muddy shores of lakes and rivers. Some propose instead that the algae arrived on land with their fungal partners already in tow. Either way, they quickly became dependent on each other. The earliest plants were little more than puddles of green tissue, with no roots or other specialised structures. Over time, they evolved coarse fleshy organs to house their fungal associates, which scavenged the soil for nutrients and water. By the time the first roots evolved, the mycorrhizal association was already some 50 million years old. Mycorrhizal fungi are the roots of all subsequent life on land. The word 'mycorrhiza' has it right. Roots (rhiza) followed fungi (mykes) into being.

Today, hundreds of millions of years later, plants have evolved thinner, faster-growing, opportunistic roots that behave more like fungi. But even these roots can't out-manoeuvre fungi when it comes to exploring the soil. Mycorrhizal hyphae are fifty times finer than the finest roots and can exceed the length of a plant's roots by as much as a hundred times. Their mycelium makes up between a third and a half of the living mass of soils. The numbers are astronomical. Globally, the total length of mycorrhizal hyphae in the top ten centimetres of soil is around half the width of our galaxy (4.5×10^{17} kilometres versus 9.5×10^{17} kilometres). If these hyphae were ironed into a flat sheet, their combined surface area would cover every inch of dry land on Earth two and a half times over. However, fungi don't stay still. Mycorrhizal hyphae die back and regrow so rapidly – between ten and sixty times per year – that over a million years their cumulative length would exceed the diameter of the known universe (4.8×10^{10} light years of hyphae, versus 9.1×10^9 light years in the known universe). Given that mycorrhizal fungi have been around for some 500 million years and aren't restricted to the top ten centimetres of soil, these figures are certainly underestimates.

In their relationship, plants and mycorrhizal fungi enact a polarity: plant shoots engage with the light and air, while the fungi and plant roots engage with the solid ground. Plants pack up light and carbon dioxide into sugars and lipids. Mycorrhizal fungi unpack nutrients bound up in rock and decomposing material. These are fungi with a dual niche: part of their life happens within the plant, part in the soil. They are stationed at the entry point of carbon into terrestrial life cycles and stitch the atmosphere into relation with the ground. To this day, mycorrhizal fungi help plants cope with drought, heat and the many other stresses life on land has presented from the very beginning, as do the symbiotic fungi that crowd into plant leaves and stems. What we call 'plants' are in fact fungi that have evolved to farm algae, and algae that have evolved to farm fungi.

The book of Isaiah in the Old Testament has it that 'all flesh is grass'. It is a logic that we might today describe as ecological: in animal bodies, grass becomes flesh. But why stop there? Grass only becomes grass when sustained by the fungi that live in its roots. Does this mean that all grass is fungus? If all grass is fungus, and all flesh is grass, does it follow that all flesh is fungus?

Maybe not all, but certainly some: mycorrhizal fungi can provide up to 80 per cent of a plant's nitrogen, and as much as 100 per cent of its phosphorus. Fungi supply other crucial nutrients to plants, such as zinc and copper. They also provide plants with water, and help them to survive drought as they've done since the earliest days of life on land. In return, plants allocate up to 30 per cent of the carbon they harvest to their mycorrhizal partners. Exactly what is taking place between a plant and a mycorrhizal fungus at any given moment depends on who's involved. There are many ways to be a plant, and many ways to be a fungus. And there are many ways to form a mycorrhizal relationship: it is a way of life that has evolved on over sixty separate occasions in different fungal lineages.

And yet mycorrhizal fungi do more than feed plants. Some describe them as keystone organisms; others prefer the term 'ecosystem engineers'. Mycorrhizal mycelium is a sticky living seam that holds soil together; remove the fungi, and the ground washes away. Mycorrhizal fungi increase the volume of water that the soil can absorb, reducing the quantity of nutrients leached out of the soil by rainfall by as much as 50 per cent. Of the carbon that is found in soils – which, remarkably, amounts to twice the amount of carbon found in plants and the atmosphere combined – a substantial proportion is bound up in tough organic compounds produced by mycorrhizal fungi. The carbon that floods into the soil through mycorrhizal channels supports intricate food webs. Besides the hundreds or thousands of metres of fungal mycelium in a teaspoon of healthy soil, there are more bacteria, protists, insects and arthropods than the number of humans who have ever lived on Earth.

Mycorrhizal fungi can increase the quality of a harvest. They can also increase the ability of crops to compete with weeds and enhance their resistance to diseases by priming plants' immune systems. They can make crops less susceptible to drought and heat, and more resistant to salinity and heavy metals. They even boost the ability of plants to fight off attacks from insect pests by stimulating the production of defensive chemicals. The list goes on: the literature is awash with examples of the benefits that mycorrhizal relationships provide to plants.

Mycorrhizal relationships have forever featured in our efforts to feed ourselves, whether we've thought about them or not. For millennia in many parts of the world, traditional agricultural practices have attended to the health of the soil, and thus supported plants' fungal relationships implicitly. But over the course of the twentieth century, our neglect has led us into trouble. In viewing soils as more or less lifeless places, industrial agricultural practices have ravaged the underground communities that sustain the life we eat. There are parallels with much of twentieth-century medical science, which considered 'germ' and 'microbe' to mean the same thing. Of course some soil organisms, like some microbes that live on your body, can cause disease. Most do quite the opposite. Disrupt the ecology of microbes that live in your gut, and your health will suffer – a growing number of human diseases are known to arise because of efforts to rid ourselves of 'germs'. Disrupt the rich ecology of microbes that live in the soil – the guts of the planet – and the health of plants too will suffer. A large study published in 2018 suggested that the 'alarming deterioration' of the health of trees across Europe was caused by a disruption of their mycorrhizal relationships, brought about by nitrogen pollution.

In 1940, Albert Howard, a founding figure in the modern organic farming movement and a passionate spokesman for mycorrhizal fungi, professed that we lacked a 'complete scientific explanation' of mycorrhizal relationships. Scientific explanations remain far from complete, but prospects for working with mycorrhizal fungi to transform agriculture and forestry and to restore barren environments have only increased as environmental crises have worsened. Mycorrhizal relationships evolved to deal with the challenges of a desolate and windswept world in the earliest days of life on land. Together, they evolved a form of agriculture, although it is not possible to say whether plants learned to farm fungi, or fungi learned to farm plants. Either way, we're faced with the challenge of altering our behaviour so that plants and fungi might better cultivate one another.

3

SELF-PORTRAIT AS A MUSHROOM IN THE DAMP AND LEAFY FOREST

Abi Palmer

I want to touch a mushroom.

I HAVE DEVELOPED such an affinity with the fungal kingdom and its fruiting bodies. To be chronically ill, to be one of *the sick ones* – a postscript in the wake of a global pandemic – is to live like fungus. No matter how dignified you appear, or how majestic, there is something of a stench of death about you. There's an association with breakdown and decay.

Like a mushroom, I appear strong and vibrant, but I'm very easily pulled apart. I have infirm, permeable edges and a lumpy, alien shape. I too am spongy and malleable. I too bruise easily. I too spend a lot of time in the dark.

Recently I have been very tired. I fall asleep late at night and wake early and unrefreshed. I am always too hot and too thirsty. I feel yeasty, fermented. My brain is very full of thoughts.

I want to connect with the earth in a thick,
damp sleep.

It's not that I want to stop thinking, but that I want my consciousness to change. It must be possible to shift away from the constant stream of internal dialogue, the worry about deadlines, about paying my bills, about whether I'm making the right decisions in life or whether I'd do better to run away and live at the seaside (and whether or not my ego would let me).

> *I want to connect with the earth in a thick,*
> *damp sleep.*

Recently I have abandoned social media. Instead, I have taken to recording long and intense voice messages, back and forth, with other chronically ill friends, whom I cannot touch and have never seen. Like me, they lie in their own little darkened rooms, like me, unwashed, like me, in various states of undress. We have sore throats and sore hands. There are months of silence followed by unbelievable bursts of productivity, where life happens all at once. We all have extraordinary talents and strange hobbies and horrible doctors and far too many financial woes.

> *I want to tap into my friends' thick, damp*
> *dreams.*

The conversations operate on Mushroom Time – refusing to manifest for long, unspecified periods, and then bursting in many directions all at once, and as if from nowhere, like pearlescent oyster mushrooms after a bout of rain.

The messages are the most transformative thing I've ever done. We talk about protest and activism and changing the world. We talk about what we can learn from the fungal kingdom: how the 'Wood Wide Web' reminds us of mutual aid networks – groups of marginalised strangers collectively passing around the same £20 to whoever needs it most that day. We talk about whether it's possible to signal and find each other the way slime cells do – whether we would want to, whether in fact we are doing it to each other right now. We record our messages at odd hours, monologuing our sleepy fungal thoughts while the world around us dreams. It feels as though we are merging and spreading, tiny spores reproducing our shapes across the airwaves.

Nobody gets to see what is really happening underneath, but these conversations are transforming me also. I am different now. I am stiller and fatter. My need to appear productive is dwindling slowly. Some of the best fungi fail to produce a fruiting body at all.

I want to carry my friends with me to the
thick, damp woods.

It is rare that I am able to access a forest, but when I can, it is by mobility scooter. I can only go as far as the battery will carry me. I spend time beforehand researching accessible routes. I am an adept rider. I handle my scooter like a horse, steering gently over impossible tree roots, and around boggy, leafy depths. Even when it panics, I am calm.

However, there must always come a time when the forest is too deep: the scooter can't go any further. At this point, if I can, I will pull out a walking stick and walk half as far as my legs will allow. My innate aversion to gravity is loosened: bouncy moss springs under my feet. I want so much to become lost in the forest, in a way that doesn't require a dropped pin on Google Maps.

I want to connect with the earth in a thick,
damp sleep.

In here, it is easier to feel healthy and alive. My lungs fill with nature, its heady top notes: all leafy foliage and singing birds. The body of the forest lies somewhere underneath all this, in the dampness and darkness and decomposing earth. I close my eyes and inhale as much of the wildness as possible, knowing that this moment may sustain me for months to come.

*I want to connect with my friends in the
thick damp earth.*

As my senses recuperate, my eyes, too, become impossibly sharp. Eventually they home in on what I'm looking for: a small mushroom clumped under a bush. And then there are more of them: mushrooms, everywhere. Milky white and delicate little buttons shoot out like beansprouts around the edge of a rotting tree trunk; mushrooms resembling other things: buns, or horns, apples and trumpets; bracket fungi fan out from tall trees, some so large a dog could sit on them.

Wherever I can, I reach out my hand. I touch gently, with a single digit. It feels so much as though I'm touching my own skin. Where the mushrooms are thick and fleshy, I think of my thighs, my stomach. The delicate papery ones are more like my hands, my bony wrists. Some are tender, others springy. Others disintegrate under my thumb.

*I want to merge with the mushroom, right
here on the forest floor.*

I wonder if the mushroom can sense my pulse, or my tenderness; whether it can sense my presence from its thick, sleepy state or whether it is wide awake and lucid, passing messages back and forth.

Touching the mushroom takes me deeper into all of it, to spaces I cannot take myself: the heart of the forest, the universe, the rooms of my friends, whom I cannot see.

I want the mushroom to touch me back.

STRANGE SOIL

Rebecca Tamás

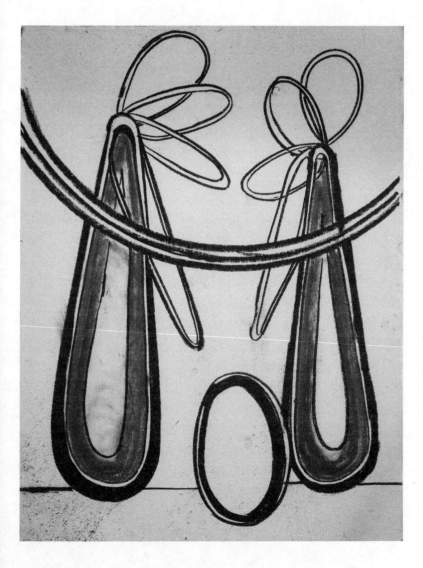

STRANGER AND STRANGER. What if we don't need a new world, but an old one? What if, hidden underneath the mud, is another life, impatient to be born? What if the very soil beneath our feet had something to tell us, a new way of living with our world and environment which has always been waiting for us?

Soil is what keeps the world alive. Soil is home to as much as a quarter of the species on our planet. Rotting leaves, fruits, plants and organisms are layered into the soil and return as something new. Good soil structure offers a wide variety of different spaces that can house organisms which, in turn, create an environment that suits them, directly altering – and improving – the structure of soil. These tiny organisms keep soils healthy and productive by passing nutrients to each other in a web of intimate, minute connections that make all life on earth possible. Complex food webs move nutrients around the system, generating healthy soils that allow human beings to eat and have clean water. The soil network, an interrelated web of organisms including archaea, bacteria, actinomycetes, fungi, algae and protozoa, and bigger soil fauna such as springtails, mites, nematodes, earthworms, ants and insects, regulates the carbon and nitrogen cycles, recycles nutrients, stores water and helps detoxify pollutants. Soil, and the many creatures which co-exist within it are the underworld which makes life possible, an ecosystem permanently in flux, always changing and moving, always creating.

Despite the reliance of all living things on soil, we don't know much about it. We do know, however, that this fragile, generative world has been damaged by intensive farming, pollution, deforestation and global heating. A third of the planet's land has been severely degraded and 24 billion tons of fertile soil are destroyed every year through intensive farming, according to the Global Land Outlook. Topsoil is where 95 per cent of the planet's food is grown and is very delicate. It takes more than 100 years to build 5mm of soil, and it can be destroyed shockingly easily. This destruction and degradation of the soil is created by intensive farming practices such as heavy mechanised soil tilling, which loosens and rips away any plant cover, leaving the soil bare. It is also caused by the overgrazing of animals, as well as forest fires and heavy construction work. These factors disturb the soil and leave it exposed to erosion from wind and water, damaging the complicated systems underneath its top layer.

Though this muddy, propagative space is what allows for human and non-human flourishing, it is being destroyed ever more speedily. We are losing good soil at an estimated 100 times faster rate than we can remake and heal it. The world's soils are thought to store approximately 15 thousand million tonnes of carbon – three times as much as all of our planet's terrestrial vegetation combined. Soils hold twice as much carbon as the atmosphere, and when soil disintegrates, the carbon is released. In the last forty years the soil in the UK's croplands lost 10 per cent of the carbon it could store. In a time of climate crisis, soil's quiet potency, its ability to store carbon safely, is utterly essential to our future survival.

Yet, for all its value, soil does not have the glamour of other types of non-human spaces or beings – it is not a dramatic 'charismatic megafauna', like a lion roaring on the plain, or a beautiful vista like a lake surrounded by mountains, or an adorable panda, or a heart-warming pet. Soil is a space that is seemingly so far from human experience, so strange, so other – its tiny networks of organisms in the dark more like an alien planet than a familiar world – that it is easy to ignore it altogether, despite our lives relying on it entirely. Soil provides the ground for all life, but it doesn't have the obvious appeal or interest of animals or plants, or the beauty of lakes, mountains and forests. Soil is what makes the verdant, irrepressible variety of the planet possible, but appears to us as quiet, blank, merely mud, ground, dirt. The subtle, hidden world of soil can be one it is hard to connect to, witness and understand. Human beings return to the soil in death, and this sense of soil as part of the underworld, of endings and finality, belies its huge energy – the energy that takes the dead, recycles them through decomposition, and brings about new life.

We know that soils are being destroyed, and that with that comes a higher risk of floods, and a more unpredictable and unreliable food and water system. An Intergovernmental Science–Policy Platform on Biodiversity and Ecosystem Services report in 2018 told us clearly that land degradation is already putting the welfare of two-fifths of humanity at risk, and that urgent action is needed to avoid further danger. There are many things that we can do to protect soils, and the organisms, plants and connections that thrive within them. Actions that can support and heal soil structure include planting 'cover crops', planting hedgerows or ley strips and encouraging the habitats of animals such as earthworms, which act as 'ecosystem engineers' and aerate the soil as they burrow into it. Using reduced till or no-till regimes in farming can also help to prevent the destruction of organic matter in the soil. Such regimes allow soil structure to remain intact, and protect the soil by allowing crop residues to stay on the surface. Better soil structure and cover support the soil's ability to absorb water, which in turn lessens soil erosion. Soil microorganisms also profit from no-till or reduced till practices, leading to a healthier soil biome.

It seems to me that because we are not telling stories of the soil, we have forgotten that, for all its seeming difference, soil is a part of us – the place in which we began and where we will end up – we have little passion for it, and therefore little energy to protect it. There is not a strong sense that a crisis in soil health is something the public is clearly aware of, or that there is a clamour for soil action within our society. There is rarely the chance for us to recognise that soil can speak, and that its capacity for every shifting-flux and creation might have something to tell us about the weary, damaged societies of the Western world. Soil is not just the ground beneath our feet, it also offers us a narrative to re-think how we might approach the strangeness of the world in which we live, and how we might re-vision our connection to nature and the land. For that reason, I want to explore two soil-shaped stories, one ancient, one relatively modern, to see what the strange soil might have to say to us, at this moment, on the edge of our choice between total destruction and new blossoming.

Antigone, the title character and protagonist of Sophocles' Ancient Greek play of 441 BC, is deeply strange. Both her brothers, Etocles and Polyneices, have died fighting on opposite sides of Thebes's civil war. The new ruler, Creon, who fought on Etocles' side, will give that brother the full burial rites, but the rebel Polyneices' body will be left uncovered, for the vultures and dogs to pick at. Anyone who disobeys this command will be put to death. But draping Polyneices in soil, in the layers of the ground from which he came, is the rite which will allow his spirit to pass to the underworld. Without returning to the soil, Polyneices will be lost.

Antigone knows that burying her brother's body in the earth will lead to her death too, but decides to do it anyway. Antigone chooses the relationship between human being, soil and death, over any chance to extend her own life. As if to underline how strange Antigone's choice is, Sophocles contrasts her with a more conventional sister, Ismene. Ismene says:

> We are only women,
> We cannot fight with men, Antigone!
> The law is strong, we must give in to the law
> In this thing, and in worse. I beg the Dead
> To forgive me, but I am helpless: I must yield
> To those in authority.

Antigone does not care that she is a woman, that she is a rebel, that she will die. She answers:

> If that is what you think,
> I should not want you, even if you asked to come.
> You have made your choice, you can be what you want
> to be.
> But I will bury him; and if I must die,
> I say that this crime is holy: I shall lie down
> With him in death, and I shall be as dear
> To him as he to me.
> It is the dead
> Not the living, who make the longest demands:
> We die for ever . . .
> You may do as you like
> Since apparently the laws of the gods mean nothing to
> you.

Antigone does find her brother's body, does cover it in soil, and is sentenced to death by an inflexible Creon. Creon follows the laws of men, which are rational and clear, and Antigone follows the laws of the gods, and the laws of nature, which are harsh, mysterious and strange. These laws take no notice of human ego or desire, but relentlessly, ruthlessly, head towards a flourishing for everything that lives.

Creon tells Antigone that she has dared to defy human law, and she responds:

> I dared.
> It was not God's proclamation. That final Justice
> That rules the world below makes no such laws.

'The world below' is the world of the soil, where the fundamental ties of living connection, of Antigone's love for her brother, overcome any momentary ideology or stricture.

The Ancient Greek word for justice is δίκη, the name of Dike, the goddess of justice and order. The word does not have the connotations that we associate it with now, but for the Greeks meant 'behaving in accordance with nature', or 'behaving in balance'. For Antigone, justice is not about power, or victory, or government. It is about returning a thing that was once alive back to the soil and dust from which it came. The soil, and Antigone's form of strange love, does not care what side each brother was on. The soil does not care what they were like, or which was kinder, funnier, better-looking, smarter; it will absorb both of them just the same, will break down their bodies into nutrients to feed back into the earth, to make new life come again. Antigone echoes this, because she does not choose between, or judge; she only cares that they were her brothers, that they were beings from the living earth, now to return. The power struggles of men and war seem paltry in the face of such soil-love, such deep-rooted figuring. This is a world in which acting in accordance with natural law is not conservative, but radical, beyond tyranny and control. This is not the right-wing rhetoric of 'blood and soil', but the risky, thrilling openness to transformation that makes all life possible. Antigone's love is revolutionarily flat, given out equally to both brothers, pressing up against the constraints of hierarchy and autocracy and female obedience. In the soil, we are all the same: merely bodies, merely water, merely cells becoming other cells.

Antigone is taken to her death, not to be executed quickly, but walled in a living tomb, away from light. The blind prophet Tiresias, who has been both woman and man, and who sees beyond the veil of the ordinary, pleads with Creon to change his mind, arguing that 'You are sick, Creon! You are deathly sick!'

But Creon revels in his power and his sickness and does nothing. Boiling under the earth, Antigone hangs herself. Her fiancé, Creon's son, finds her corpse and then stabs himself. This is followed by the death of Eurydice, Creon's wife, who commits suicide when she finds out what her son has done. Creon's entire family is destroyed by his choice to ignore the justice that calls to us from the dust, from the curling, mute mouth of the soil. Creon, at the very end of the play, recognises this:

Whatever my hands have touched has come
    to nothing.
Fate has brought all my pride to a thought of dust.

Antigone, strange rebel, wanted only to restore balance, and yet was seen as a terrible threat to the status quo.

I know that if my brother were to die, I would risk my life to see his body in the ground, without shame, safe. I know that that makes no apparent sense in a world dedicated to the appearance of comfort and safety. Still, it is true.

To act with a rational confidence that the dead are gone is to forget that their bodies, their energy, break down and return to the earth on which we feed. They are in our teeth, our bones, our blood, and will we do nothing to reckon with them? Dust, and grit, and feelings which we cannot fully explain, that come from the irrational, hard-to-control parts of ourselves, the animal parts that call to us of the communion between beings.

Strange, or not strange, that when you love things, like water, like air, like soil, you care for them. Strange, or not strange, that when you care for those things, the dead stay peacefully where they belong, and you survive. Antigone's story shows us that our returning to the soil is not a comforting metaphor, but an essential, invaluable right, whether that be bodies in the earth, ashes in the sea, bones pulverised and floating into air, to rest again in new turf. To return to the soil is to recognise that our human bodies are only one part of our cycle, and that we belong to this strange turning of the world's wheel, the underworld where we are re-made into new shapes, to come again into the light.

Once, someone told me about how Mussolini ordered the graves of soldiers from the First World War to be opened, and moved their bones to a new and towering Fascist tomb, celebrating Italian military might. Why did that information make bile rise in my throat, as if I might be sick? These dead men, ripped from the earth, bodies made to say *power*, dragged from the justice of the ground. The tomb was white, the colour of frozen death, of suffocating purity. The empty soil was black, lamenting. Antigone knows the earth is full of electricity; that it is not still, like monuments, palaces and tombs are, but jittering, alive. The story of the soil is the story of how all of us are connected, owing intense fidelity to each other's difference, and to the intricate balance that allows all of us, human *and* non-human, to thrive.

Despite being written thousands of years after Sophocles' play, *Penda's Fen*, a BBC Television 'Play for Today' written in the 1970s by David Rudkin, also sparkles with a strange, soil-born liveliness. It takes its lineage from the rich 'folk horror' tradition of *The Wicker Man* and *The Witchfinder*, but stands strangely alone – not true horror, but a film of unsettling beauty, hope and mysticism. The film follows another unexpected rebel and journeyer of the soil: Stephen, son of a vicar, living deep in rural England. Stephen struggles to move past his own outwardly rigid conformity and faith, and to accept his burgeoning homosexuality and vivid, fragmented identity. The soil here is more metaphorical than Antigone's, but no less potent – Stephen must battle to find the true meaning of his local ground, to offer up freedom rather than conservatism, to take back his inheritance from the forces of control and repression.

Stephen is not like the other boys in his patriotic, macho school in the Malvern Hills – he is strange, dreamy, intelligent and suffused with complicated desires. He has visions of demons and angels fighting on the church spire, talks to the ghost of Edward Elgar in a broken-down shed, and is painfully drawn to the radical leftist playwright next door, and his talk of conspiracies and cover-ups. Stephen's rural idyll is not the cliché 'chocolate box' version of perfection it seems on the surface; rather it is a battleground where the forces of conformity, control and authority wrestle the soil-deep forces of freedom, joy, openness, difference and revolution.

On Stephen's eighteenth birthday, he finds out he is adopted, and must confront the fact that he has become strange to himself. His seemingly quiet and conventional adopted father reveals that he is, in fact, a political radical – believing Jesus to be an 'elemental village god', the force of life and disobedience, not of control and law. This Christ is a pagan one, in the old meaning of pagan, 'of the village'. A local Christ, a revolutionary Christ, a Christ of the soil, 'by whom the earth has been haunted since the first beat of the heart of man'. Christ here is not a remote figure of power and distance, but a local, intimate one, of field and wood, heart and hearth. A Christ who empowers human beings to extend their joy and tolerance and questionings, not suppress them.

Stephen begins to grow into this new fractured reality, of pagan Christian gods, of strange parentage, of sexual pleasure, of landscapes containing the secrets of revolution, freedom and desire. Like Antigone, Stephen is fated to restore the elemental forces of earth, of balance, of soil and of movement. As his internal self develops and mutates, so signs in the outward world begin to hint at his hidden inheritance. His village's name, Pinvin, begins to change on signs and roadmaps, to 'Pinfin', and then into 'Penda's Fen' – Penda being the name of England's last pagan king, who fought to save his earth-rooted, soil-rooted way of life, on the very ground Stephen walks on. As Stephen's father wonders, when thinking of Penda's death, 'what mystery in this land went down forever [when he was defeated by a rigid Church] . . . what darkened sun of light went out?'

But *Penda's Fen*'s look back to the clouded ancient past is not a harking back to a perfected, whole or secure nationhood or identity. It is a revelling in the strange, the atonal, the different; the openness of growing and dying things, of wild revels in the woods, of fractured selves, their coherence muddled beyond repair. Penda, the King of Malvern's fertile soil, is 'our sacred demon of ungovernableness'. His land has room for imperfection, difference, freedom – it is the antithesis of a *Country Life*, conservative immobility; of duck ponds and Astroturf; of choked down tea and scones, of twitching net curtains, little England prejudice and cruel judgement. When Stephen is attacked by the forces of puritanical Christianity and morality, he rebels, saying, 'I am nothing pure. My race is mixed. My sex is mixed. I am woman and man. Light with darkness. Mixed. Mixed. I am nothing special. Nothing pure. I am mud and flame.'

He calls on the spirit of Penda, who arrives in a glorious burst of green and yellow fire from the burning mud, to protect his strange inheritor. Inheritor of mixed earth, mixed land, disturbed soil, strange hills and rivers and light. Penda gives one luminous entreaty to Stephen: '*Be secret. Child, be strange, dark, true, impure, and dissonant. Cherish our flame.*'

To my mind, Penda is asking Stephen to take after the true, mixed nature of the soil – its merging of different plants and organisms, its balance between what is rotten and dying and what is newly born, its linkage of difference into a paradoxically cohesive dissonance. Penda's flame is one that revolts against Olde Worlde England, against fixed histories, nationalism, racial purity, sexual purity and violent morality – stories which try and make the soil speak of limits, restraint and hierarchy. In *Penda's Fen*, the freedom to change and be changed, through the occult vitality of the landscape, is more powerful than any of those. Stephen's home, which at first seems to be a pastoral, quiet and conformist place, is revealed to be a site of revolution, where white turns green and brown, and disobedience is the path to happiness. This uncanny environment vomits up the strange history of mystics and outlaws, the history that sits under the surface of polite fields and inked-in family trees, inside the explosive soil – bursting with desire, with room for the human and non-human alike to dance their odd and ever shifting steps.

If the old world has anything to give the new one, it is this submerged vision that Antigone and Stephen share, thousands of years apart, of what could be: pagan disobedience, love made from the ground. Soil, and its network of varied, diverse, life-creating organisms and intimate connections, is the space in which we become possible. We need to tell stories like *Antigone*, like *Penda's Fen*, to remind us that our connection with strangeness is not something to be disturbed by, but something to embrace. The plants and creatures that make up the soil, in their own way, give birth to us, and take us home again at the end of our lives. They tell us that we are not the rigid, lonely monads that capitalist society tells us that we are – but that we come from, and are made of, eternal, ever-moving connections, a series of linked, evolving points, in glorious, collaborative balance with each other. Despite the world of the soil seeming deeply alien and unfamiliar, it is our world too, part of the intimate links between us and the nonhuman. The soil, as it feeds and supports the food that we eat and the environments we live in, creates our life; and the soil, as it absorbs our body or our ashes back into the mud, creates our death too, filtering us back as things transformed, as pure energy, ready to begin again.

That soil story, of connection and interdependence, of flat hierarchies, of mutual transformation and networked flourishing, is in itself the story of an ecological life, and it is the story we need to tell over and over again to find a new environmental balance, and a future which can sustain every single one of us.

NOTES

31 'These factors disturb the soil . . .': www.nhm.ac.uk/discover/soil-degradation.html.
'The world's soils are thought . . .': www.checinternational.org/the-importance-of-soil-on-life/.
'In the last forty years . . .': www.theguardian.com/environment/2021/apr/16/poor-mans-rainforest-stop-treating-soil-like-dirt-aoe.

'An Intergovernmental Science–Policy Platform . . .': www.
ipbes.net.
'Using reduced till or no-till regimes . . .': royalsociety.org/
topics-policy/projects/soil-structure-and-its-benefits.

33 'Soil microorganisms also profit . . .': regenerationinternational.
org/2018/06/24/no-till-farming.

34–8 'We are only women . . .'; 'If that is what you think . . .';
'I dared . . .'; 'Whatever my hands . . .' All quotations from
Sophocles, *Antigone*, translated by Dudley Fitts and Robert
Fitzgerald (Harcourt, Brace, 1939).

40–42 'an elemental village god . . . *Cherish our flame*': All
quotations from *Penda's Fen*, written by David Rudkin, directed
by Alan Clarke (BBC Television, year of transmission 1974).

5

PLANTS KNOW

Emanuele Coccia

PLANTS PERCEIVE THE WORLD in a much more refined way than we do: they have more senses than we do, and their sensitive world is not simply divided into five great realms. They have memory, and make this memory their own body: a tree is the climatic archive of the years in which it lived. They interact with their fellow plants to communicate the arrival of enemies, including individuals belonging not only to different species but also to different kingdoms: flowers are some of the most extraordinary tools of inter-kingdom communication that have ever been invented. *Plants know.* They know perfectly well what happens around them. They know perfectly well what happens inside them. They know perfectly well the difference between the outside world and their body; they feel the frontier between the 'I' and the 'non-I'. Every plant is self-conscious, and yet technically it doesn't have a brain, at least not in the anatomical sense of the word. Plants can simply 'perform' activities like thinking, perceiving, knowing and perceiving through their somatic equipment.

In animals, each function is performed by a specific part of the body, whose form becomes specialised and differentiated: what we refer to as organs. In plants, the functions are taken over or performed in a widespread and plural way. For instance, regarding perception, there is not a single part of the plant's body that feels the outside world following a specific dimension (as is the case for the five senses): the perceptual function is spread throughout the body, through non-specialised sensors. Hence it takes place on a molecular basis. Intelligence and the plant's problem-solving abilities are not concentrated in a single portion of the body (as with the brain), but instead are spread over its entire extension, and therefore intrinsically multiplied. The plant is by definition without organs, and the concentration and definitive specialist differentiation of tissues, the nature of the plant's body, is almost completely absent or otherwise involved in a dynamic process in which what has been differentiated can be de-differentiated and return to being what it was.

In other cases, plants differentiate the part which can perform a particular function by producing many examples of them. Take the basic function of reproduction: a plant is a living being that can build hundreds of sexual organs (imagine if you were to have sex not only with a penis and a vagina but with 150 of them, possibly all together, with different individuals).

The reason for this anti-organic logic is, as botany has long claimed, the sedentary character: a plant cannot move and is exposed much more often, and much more intensely, than an animal to the attack of predators. Precisely for this reason, it cannot concentrate a central function in a single part of its anatomy, but must entrust it to several parts, in a widespread and equal manner. A plant, if you will, is an organism that never ceases to build plan Bs, second bodies, alternative anatomies. It's like growing hundreds of noses when faced with the danger of losing one to a predator, or developing lungs almost everywhere in case a lion or vulture comes along and rips one set to pieces. It's for that reason that, as botanists have demonstrated, the somatic construction of the plant takes place in a modular way and through the reiteration of simple formal units.

We should imagine the same structure for what we call consciousness, or 'I'. A plant is not a being without a 'me', it's a being whose 'me' is reflected in hundreds of portions of the body. Each plant is an individuum with a non-pathological form of schizophrenia. Every plant, from this point of view, is not only a self but is much more of an 'I' than I am, or we all are. It doesn't say 'I' once, it says it hundreds of times, simultaneously in the same body, in different forms, exactly as every plant does anything else in hundreds of different places (be it sex, photosynthesis, or anything else). The plant is capable of saying 'me' with any part of its body.

They say 'me' more intensely, more decisively, in the same way as they have sex much more intensely than we do; they breathe much more intensely than we do, they feed much more intensely than we do (they have hundreds of stomata for absorbing carbon dioxide and hundreds of pores for capturing water).

From this point of view, the self is also, and above all, a vegetable: fact. To be an 'I' we do not need to have a brain, sense organs, eyes, ears, nose. To have an 'I', it's enough to have a living body, and a body whose main characteristic is the fact of being born. Here's the point: if we ask plants to explain to us what an 'I' is, their answer is that the 'I' is originally and constitutionally decerebrated, without brain, and without organs, but remains the main plastic force of a living body. The 'I' is the property of a body (there is no 'I' without a body) capable of growing and shaping itself. The 'I' is the property of a born body.

Let's attempt to revisit this in a slightly more technical way: the 'I' has always been the form of reflexivity, the way in which something is able to bend and return to itself. Even in saying 'I', we create with our breath the language, the sound, the word, only to capture ourselves saying it. The 'I' is the first circle of the living being. Now, the first form of reflexivity is not that given by sensation (eyes, nose, ears, etc.), or through brain intellection. The first form of reflexivity is that which is applied as we grow. It is growth, taking shape and having a form that defines the being, the ego. It's only because we have to grow that we have an 'I'. Which means that we are an 'I', we have an 'I', just because we are born. And it doesn't matter if we're an orchid, an oak tree, a dog, a rabbit. Everything that lives is born and is an 'I'.

What makes each of you an 'I' – someone or something that needs to distinguish itself from the rest of the living and to affirm itself as different – is birth itself.

To question oneself about the metaphysical structure of the 'I' means to question oneself about the form, nature and sense of birth. In short, the science of the ego, psychology or egology, is the science of birth, or rather of nature, because nature, as we often forget, is the way of being of all that is born. It is only by asking ourselves about birth that we will understand why we have an 'I', and why this 'I' is a universal fact, the most basic form, elementary, even if not spontaneous, of life. There is no life which cannot say 'me'. There is no life that is not me.

6

WILDER FLOWERS

Rowan Hisayo Buchanan

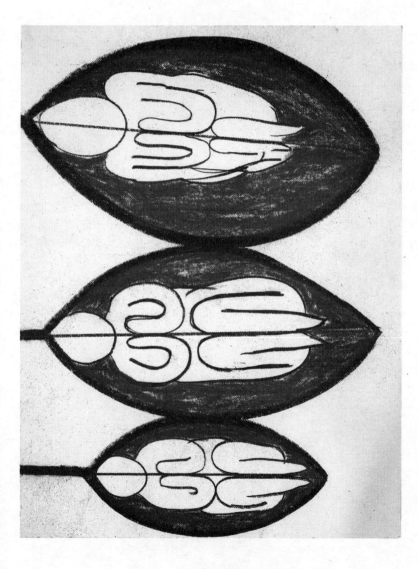

CAN YOU REMEMBER the first time you drew a tree? Perhaps you crayoned a lollipop – a brown trunk topped by a green blob. Perhaps you aimed higher – a cloud on a stick.

I have a memory of looking out of my schoolroom window. The glass offered a slice of our closest tree. The crown reached up and out of view, but I could see the trunk and a criss-cross of branches. It was winter. The London plane trees were bare and their trunks blotched and scabby. I liked to watch the way the thinnest twigs danced. I can't remember if I'd been told to draw a tree or if the subject was my own choice. What I do remember is the knowledge that if I tried to match its detail, I'd only make a mess. I was at the age when most children are conformists, and I had a fear of the ugly. Meticulously, I drew a Sprite-green lollipop. It was more important to me that my picture be pretty than that it be true.

This is the story of an attempt to learn to see our green world. Landscape painting has always had a wobbly reputation. The startling saturation of Monet's landscapes has become associated with fridge magnets and peeling mouse mats. Van Gogh's *Starry Night*, despite its genius, is attended by a whiff of dentists' waiting rooms and dormitory bedrooms. Back as far as 1669, the art historian André Félibien, on behalf of the French Academy, ranked the genres in order of their merit: history painting, portrait painting, genre painting, landscape painting, animal painting, still life. I will save a defence of the last two for another day. Today, I'd like to talk about landscapes.

For years, stuck to our fridge was one of my first non-lollipop trees. It was done with charcoal on thin, greyish paper. A tree drawing made with a burnt tree on a pulped tree. My mother loved that picture; she said it had tree-ness. I think she hoped for a forest. I wasn't interested.

I scribbled down humans instead: people I'd met and people I'd never met. Girls with flared sleeves or flared trousers. And sometimes bald aliens with long, tilted eyes. These drawings were not particularly good. I'm not sure one ever ended up on the fridge.

But eventually, I realised that people could not stand alone in white voids. A person without a place was barely a person. And so, I went back to landscape and to trees.

Some things I learned:

1. A tree is an endless splitting. Trunk to branch, to slimmer branch, to skinny twig, to leaf vein. Each division will be thinner than the last.

2. Leaves have direction. Some, like those on the Judas tree or horse chestnut, hang down. The olive's leaves start up in all directions, like the tufted hair of a schoolboy who has just emerged from an impromptu brawl.

3. The light does not just fall on trees. It falls through them. Turning the world greener underneath each leaf.

I learned these rules from books and teachers and from sitting with the dew soaking through my jeans and my sharpener getting endlessly lost in the grass blades. The more I looked, the more there was to see. If you rely on your assumptions, you usually get it wrong. That's why novice artists always draw the eyes too far up the face – you draw what you think is there, not what is. The battle is to see the tree and not the lollipop.

I needed to unlearn even the rules I taught myself.

What I said about trees getting thinner and thinner – not always true. There's a cherry tree on a street I walk past every day. The trunk is slim and narrow, the bark a smooth papery texture. As you follow your eyes up the trunk you see that it gets wider, nubblier, curvier. It has been crown-lifted. This means the bottom branches have been sliced off in order to raise the overall head, or crown of the tree. That's what left those scar knobbles. It's often done in order to stop a tree blocking light, though in this case I suspect it's so you aren't smacked in the head as you proceed down the street staring at your phone or your dog. It's a slightly gentler version of what is often done to London plane trees. When you look up and it appears the tree is raising thick fists to the sky, that's a pollard.

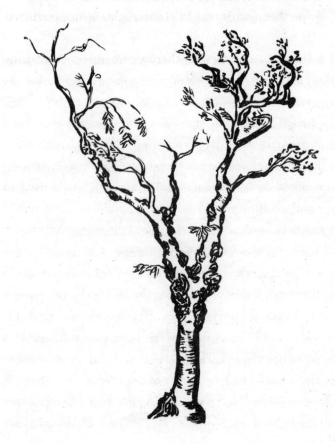

Tom Dickerson, on behalf of the Royal Horticultural Society, lists three reasons to pollard:

1. Preventing trees and shrubs outgrowing their allotted space.
2. To reduce the shade cast.
3. To prevent electric wires and streetlights being obstructed.

His article is careful to state there are 'many other' reasons, and a quick search tells me it might sometimes be for the health of the tree.

I realised that when you draw trees, or grass, or flowers, you're drawing people too. That is true whether what you sketch is carefully planted or a sneaky weed. Along my London walks, I often see small purple plants. I used to think they were clover. I didn't check their true name until I stopped to draw them. No, they are oxalis, otherwise known as creeping wood sorrel or sleeping beauty. They self-seed in walls and cracks in the pavement. Wild as can be. And yet they thrive in gaps human hands left.

I used to have a pocket of my mind called *nature* – a green-brown blur. But I've learned that drawing a plant is almost always drawing our interventions, our cutting, our pruning, our scarring, our reshaping of the land. I hate to disagree with old André Félibien, but landscape painting is a form of history painting.

A few years ago, I went to a National Gallery show of the work of George Shaw. Shaw is a contemporary British landscape painter. And not just any landscape. His subject is the Coventry housing estate he grew up in and the surrounding streets and woods. This show was a show called *My Back to Nature*. Punny, right? On large canvases in enamel paints, he rendered scarred trees, spotted by spray paint. He gave us old porn mags lying scattered on the forest floor poking out between the ivy. Crumpled cans collected inside the hollow of a split trunk. An old blue bin bag mostly disappeared under browned leaves. Both he and the National Gallery's publication were quick to draw parallels to classical art: the trees' wounds were related to Jesus' wounds, the blue bin bag might be the Virgin Mary's blue cloak. They noted that he studied the great masterworks of the gallery. They offered up Piero Del Pollaiulo's *Apollo and Daphne*, in which Daphne's stretched arms are transforming into branches. I trust this is true. But when I remember going to the gallery and standing there under the blue green leaves and looking at the ridged trunks of the trees, I remember a smell. It is a wet, cold odour

– a perfume of mulchy sweetness. It could not possibly have been in the National Gallery. That is exactly the sort of smell that they have experts to avoid. No one wants *The Madonna on the Rocks* to rot. Yet there was something about Shaw's trees, the girth of them, the muddy dapple, that sang so true that I cannot remember them without the smell of the woods. The cans, the plastic bag, the abandoned mattress, the penis scrawled on a trunk – each of these feels true to the sweet, dark woods of my memories. He is a man who sees how we collide with trees.

If there was a spell I could cast to draw something that smelled as green as Shaw's work, I would sing it. Instead, I draw slowly and I plant, trying to summon the green closer. We don't have a garden where I live with my partner, but we do have pots. We have an olive tree with a head almost as round as a lollipop. Looking at it makes me think of how Van Gogh complained of the difficulty of painting olive trees. He wrote to his brother, listing all the colours he saw in them – *old silver, sometimes with more blue in them, sometimes greenish, bronzed, fading white above a soil which is yellow, pink, violet tinted orange.* Next to the olive in its own pot is a shrub daphne with white, star-like flowers. I'm always struck that it shares its name with the girl who turned into a tree, who has been painted and sculpted so many times.

Recently, still dizzy from the lifting of lockdown, my partner and I found ourselves poking our noses into shops. Curious to see new things not backlit by a screen, we wandered into a Japanese print specialist. For a moment, I held onto a yolk-yellow print of a marigold. They aren't native to Japan, but this print was only a few decades old. Marigolds had made their way east and west across the sea, carried by human hands from the Americas. I loved it. My partner wasn't sure. We slid it back.

The next day, wondering if we should have bought the print, I remembered something I'd read. In the Japanese Middle Ages, *karesansui* gardens were planted to look like the ink paintings of wild nature. Plant to painting to plant. Why shouldn't I do the same? At home, I ordered a tray of marigold plug plants and, because they were on sale, another of apricot snapdragons. They're two weeks old now – small and green. No sign of gold, but I am hoping. I know how lucky I am to live in the city and have this small, green gallery of pots.

This melding of art and land isn't unique to Japan. Indeed, the first use of *landscape* recorded by the *Oxford English Dictionary* refers not to dirt and to trees but to the making of a picture of the land. And it is not until nearly a hundred years later that it records the use to mean the dirt, the land, itself. Landscape was to land as portrait is to person. When we talk about the British landscape, the urban landscape, the rural landscape, we are talking about art. And as humans, well, we tend to think we can improve upon our art.

Uvedale Price, as well as being the bearer of a rather remarkable first name, was a successful landscape designer at the end of the eighteenth and the beginning of the nineteenth century. It was an era of technological progress, but also the era of Percy Shelley wandering around looking windswept, Turner painting crashing seascapes, and William Blake writing incendiary poems. In the spirit of these times, Uvedale Price was one of the men imagining a wilder kind of garden. He was a member of the Picturesque movement. When you see the word used in travel brochures, *picturesque* is the old-school version of *Instagrammable*. Something so pretty that it is like a picture. For Uvedale Price and his peers, it meant that their landscape design was inspired by art. He even went so far as to tell would-be landscapers that they must study paintings – must look not only at nature but also at nature as seen by artists. This would guide them as to how they might *improve* the landscape. He and his allies are often called the *Improvers*.

When Uvedale Price told his followers to look at paintings he didn't mean Raphael's ordered gardens. No. To him the Picturesque described a very specific sort of appeal. He was a fan of the French artist Claude Lorraine, known for his wild, ruin-dotted landscapes often framed by wonky-trunked trees. He was so popular that artists and landscapers took to carrying around little curved Claude mirrors to give the world the same soft painterly appearance as his paintings. I'm reminded of the way the world seems to people more beautiful on Instagram than it does in the physical plane.

There is no way for me to visit Uvedale Price's home garden, Foxdale. It's in private hands, and in any case the still world feels only cautiously ajar. So instead, I click through image after image of Lorraine's from the British Museum's digital archive.

I pause on a sketch. The description says 'Pine tree'. The inking is fine and careful. He's scratched in brown ink over graphite. In places, the leaves are only a pen dash, but in others he's made short, jagged marks. His pen tells me that the trunk is crooked and rough. It is hard to know how long this tree would have taken to draw, but there are maybe a thousand small brown marks. I imagine him moving his hand up and down for each flick and dab and dash of ink. This is only a study, and yet it is scratched by detail. This, I am sure, is a man who wanted dearly to know that tree, to get the memory of it into his muscles.

Under the tree are two figures, sketched in pencil. Faceless, armless, there as if only for scale. I can't help but smile to see the tree being star of the show. This piece, along with many others, was donated by Price's Improver friend Richard Payne Knight.

Uvedale Price tried to explain what exactly it was about the paintings he loved that he was trying to capture, when he called something picturesque. He claimed they had an appeal opposite to that of the neatly beautiful. It was landscape that had 'roughness' and 'sudden variation'. For example, the large, exposed roots of the beech tree, ruins, and wild plants and creepers. When he suggests that 'sedums, wall-flowers, and other vegetables that bear drought, find nourishment in the decayed cement,' he might as well be describing a garden for some new Shoreditch development. There's something funny and familiar about the idea of 'improving' the landscape to make it look more ruined.

Still, who am I to laugh? I am just as guilty of trying to grow an image I've loved. Hunting after that single gold bloom, I was now responsible for a fleet of marigold plugs. Slowly, I poked them out of their plastic tray and into larger pots. The task was fiddly and precise. I tried not to tear the white nests of roots. To keep me company, I put on a radio programme in the background, the popular *Gardeners' Question Time*. A caller asked about the possibility of creating a wildflower meadow. Christine Walkden, a horticulturalist and 'plants-woman', explained that wildflower meadows, though 'great', are also a great deal of work, and that calling wildflower meadows a 'panacea' is 'total, utter bunkum'. The other hosts agreed that a wildflower meadow might be more work than it is worth. Each offered an alternative. James Wong, an ethnobotanist and garden designer, said a moss garden might be more appropri-ate to the British climate than a wildflower meadow. Matthew Wilson, a landscape designer and broadcaster, said that many of his clients were asking for wildflower meadows. And, while not entirely unsupportive, he bemoaned 'lawn shaming'.

Only a few days later, I met an old friend, a balcony gardener herself. She told me with excitement that she and her partner had ordered wildflower seeds. They had just planted them and were looking forward to sprouts. I asked what flowers. She wasn't sure, only that they were wild. I didn't have the heart to tell her about *Gardeners' Question Time*. It was May, but it was raining, and as we walked water fell down our necks.

Was the unwildness of the wildflowers an excuse to lapse into tameness? Were marigolds really the right things to be planting? I had been distracted by beauty.

Perhaps a nineteenth-century Improver would have found the mirage of wildness to be enough. But these days, we're all conscious that we're losing our wildernesses. As I learned so many years ago, we cannot stand in white voids. We cannot survive without our lands.

There are those proposing a new wildness. In *Rambunctious Garden* Emma Marris has written about how we must learn to celebrate the small and local wildernesses, and new wildernesses that aren't premised on the exclusion of humans but rather coexistence. When Uvedale Price declared his love of ruins, he was simply trying to grow a beautiful scene. But as the world veers towards ecological chaos, if the only wild worlds we can imagine are ruins, then we condemn ourselves to be ruined.

To draw a plant or a landscape is to say what the world is, but also to suggest a vision for what the world might become. I try to imagine a wilder world that is also a human world. I can't quite see it. In her book, Marris suggests a layered solution: one in which we have many different goals – the protection of particular species, the promotion of the rights of animals, the slowing of extinction, the promotion of biodiversity, genetic diversity, the experience of nature – and we must juggle all of these and more. As I read, something in my brain flickers. I catch at the edges of an image of what it might look like. I imagine James Wong's moss garden. I think of how Marris celebrates abandoned spaces. I consider the nettle patch thriving by a block of flats I often walk past. The nettles have taken over the whole front and grown as tall as my face. Something in me complains. Do I really want nettles? I don't like them. When the wind picks up their thin leaves, I flinch. Too many times when I was a child, nettles raised pink pockmarks along my arms. I scold myself at my resistance. Nettles aren't as delectable as what we more typically call wildflowers, but they feed ladybirds and caterpillars, which in their turn nourish hedgehogs and amphibians. (Determined humans can make tea or soup out of them.) I tell myself we must allow the nettles, and comfort myself that there will be wildflowers that are easier to love – the bee orchid, the poppy, the violet. I daydream of different parks, gardens, forests doing a thousand different things. If plants can become art can become plants, then I'm starting to think I should consider what I draw.

Last year, the sky on the west coast of America turned grey and orange because fires were ravaging its forests. I teach online, and some of my students were in California. One of them gestured at his window and said in not so many words that it was hard to write a story when his sky was on fire. I think a lot about this challenge. I ask myself how to make art in a world that is so often burning. But we need art that tells us what is on fire, and why, and how it might be different. If you can draw it, or photograph it, or render it, or write it convincingly enough, then you may summon new improvers with a new dream of wildness. And this time they may be able to improve the world for us all.

I'm going to start small. If I can convince no one else of the nettle's charm, perhaps I can persuade myself. In an early draft of this essay, I wrote, *We need to learn to look without a filter or curved Claude mirror.* But I think I was wrong. Why not start with beauty and seduce ourselves on nature's behalf? So today, I will head out with pencil and ink to learn to love the nettle. To try to find its grace.

SOME WORKS CONSULTED

Mavis Batey, 'The Picturesque: An Overview', *Garden History* 22, no. 2 (Winter, 1994), pp. 121–32; www.jstor.org/stable/1587022.

Tom Dickerson, 'Coppicing and Pollarding', RHS *The Garden*, January 2013; www.rhs.org.uk/about-the-rhs/publications/the-garden/the-garden-back-issues/2013-issues/january/How-to-pollard-and-coppice-shrubs.pdf.

Hélèna Dove and Harry Adès, *The Botanical City: A Busy Person's Guide to the Wondrous Plants to Find, Eat and Grow in the City* (Hoxton Mini Press, 2020).

Jessica J. Lee, *Of Field and Forest: Aesthetics and the Nonhuman on Hampstead Heath* (York University, 2016); yorkspace.library.yorku.ca/xmlui/handle/10315/32682.

Claud Lorrain, drawing: www.britishmuseum.org/collection/object/P_Oo-7-179.

Emma Marris, *Rambunctious Garden* (Bloomsbury USA, 2011).

Markus Poetzsch, 'Vying for "Brilliant Landscapes": Claude Mirrors, Wordsworth, and Poetic Vision', *The Wordsworth Circle* 39, no. 3 (Summer 2008).

Uvedale Price, *An Essay on the Picturesque, as Compared with the Sublime and the Beautiful and, on the Use of Studying Pictures, for the Purpose of Improving Real Landscape* (Wentworth Press, 2019).

Sakuteiki, *Visions of the Japanese Garden*, translated and annotated by Jiro Takei and Marc P. Keane (Tuttle, 2008).

George Shaw, *My Back to Nature* (National Gallery Publications, 2016).

Tom Williamson, *Humphry Repton: Landscape Design in an Age of Revolution* (Reaktion, 2020).

Norbert Wolf, *Landscape Painting* (Taschen, 2017).

Gardeners' Question Time at Home: 'A Midsummer Night's Dream' (BBC Radio 4, first broadcast 23 April 2020).

National Gallery of Art: www.nga.gov/features/slideshows/vincent-van-gogh.html#slide_4.

7

BITTER BARKS

LEGACIES AND FUTURES OF THE FAMOUS FEVER TREE

Kim Walker & Nataly Allasi Canales

Bajo qué huaca oculta, este país. En qué color de piel, su marcha hacia ninguna parte. Qué aguas flamenco y zorro beben del mismo pozo. Sobre el río viaja el indio en su canoa. Árbol de la quina, tus hojas cubren nuestra falta. Pronuncia nuestro nombre. Birú Perú. No lo reconocemos. Cuánta nada hemos construido. Cuántos huaycos de palabras, como niños aprendiendo a escribir.

(Under what *huaca* hides, this country. In what skin colour, its march to nowhere. What waters flamingo and fox drink from the same well. On the river the Indian travels in his canoe. Quinine tree, your leaves cover our lack. Speak our name. *Birú* Peru. We do not recognise it. How much nothingness have we built. How many landslides of words, like children learning to write.)

Teresa Orbegoso, 'Perú', *Poemas de Teresa Orbegoso*, translated by Nataly Allasi Canales (2016)

QUINO, *ÁRBOL DE CALENTURAS*, fever tree, *Cinchona*.

This is a plant that has been given many names over the centuries, written about countless times, and used by different people for different purposes: local healing tree, miracle cure, imperial tool, national symbol. From the 1600s onwards, the bitter bark of the tree was the only known effective cure for malaria and its fevers, and was so valuable it now resides at the centre of the Peruvian flag. At the Royal Botanic Gardens in Kew, preserved on the museum shelves, are 'miniature forests': thousands of historic *Cinchona* herbarium, seeds and bark collections. Large collections of *Cinchona* specimens can be found in many botanical and pharmaceutical institutes around the world, representing the interest Europeans had in obtaining it to understand, analyse and control its elusive chemical secrets. But it was also an important tree to the Indigenous peoples it belonged to. The people of South America, like many Indigenous peoples around the world, have had the power over their land, biological wealth and knowledge extracted unjustly, and still suffer the repercussions today. We are two researchers who have spent the last three years exploring the history, nature and relationships of this beautiful tree with its environment and the communities it touched over time and space. We have been thinking about these collections and their meanings in a decolonial setting as part of an initiative at Kew. Here we share some of our thoughts on this plant, on how its histories inform the past, and how the powerful compounds it contains still impact upon the present, not least during the beginnings of the recent pandemic.

Humans have been hacking plant chemistry for their own use as long as they have been around. One group of compounds, alkaloids, are an important category of medicinal drugs, and include morphine, caffeine, cocaine and quinine. It's this last chemical that is sourced from the bitter bark of the *Cinchona* tree, and was one of the earliest chemicals extracted in the newly rising chemical industry of the early 1800s. In May 2020, Donald Trump claimed that hydroxychloroquine was an effective coronavirus cure. From 1945 hydroxychloroquine replaced quinine as the main malaria treatment, and ever since it has repeatedly and mistakenly been described as a synthetic quinine. It isn't. It is merely a synthetic antimalarial, and a different chemical altogether. But during the Covid pandemic people were desperate for miracle cures and, since hydroxychloroquine is difficult to get hold of, social media posts began popping up describing *Cinchona* bark as a 'natural' cure for Covid, even though *Cinchona*/quinine's harmful side effects and potential toxicity mean using it as a medicine can be dangerous. As people began to stockpile the bark and other sources of quinine, a large international tonic water company released a statement about the fizzy, quinine-containing soda, making it clear it was not a cure for the virus. This time, at least, no miracle cure was to be had from *Cinchona*'s bark. But by then rumours of hydroxychloroquine's, and by extension quinine's, efficacy against Covid had reached the original *Cinchona*-containing forests of South America, where Indigenous groups, already at higher risk of death from it, started to look at *Cinchona* bark as a potential prevention or treatment. When Spanish colonists invaded these forests, Indigenous peoples were tragically depleted by diseases. Later, white Europeans appropriated their medicinal plants. Now, once again, a disease, the fever tree and Indigenous and non-indigenous humans have become entangled in complicated ways.

The fever tree symbolises our continuing efforts to exploit and subdue nature; in the process it has become an instrument of power. Quinine was used to enable Western expansion into tropical regions across Asia and Africa, through its ability to protect the health of both colonialists and the indentured Indigenous labour force working for them. *Cinchona* and quinine came to be controlled by Westerners – as an imperial tool. In the nineteenth century, British and Dutch colonialists began to fear that their supply of *Cinchona* might be threatened by overharvesting and political instability in South America (the latter as a result of Spanish colonialism), and so *Cinchona* was taken from its original cloud forest home and transported across oceans to British-Indian and Dutch-Indonesian plantations. Once its seeds were being cultivated on these competing plantations, it became more difficult for the South American communities to trade in the bark. Moreover, to produce quinine in mass quantities, natural vegetation was burned to make space for monoculture plantations.

The translocation of the tree had a further impact on Indigenous land use, through the drafting in of local, and non-local, indentured labour in India and Indonesia to work the plantations, which caused ripples biologically, culturally and socially. Although the British claimed to be driven by a humanitarian impulse to make a cure for malaria available to everyone in India, ironically the change in land use, combined with the digging of drainage ditches created for roads and railways, only helped spread malaria faster. On top of this, the particular strain of fever trees they cultivated were not necessarily high in alkaloids, thus falling further short of the aim of mass-producing a cheap fever remedy. Instead, it was the Dutch who went on to lead the global trade in quinine and establish one of the first pharmaceutical cartels to control its availability.

It is now illegal to take a plant, and the knowledge of its use, from one area of the world to another for profit, or indeed any other reason. Developments such as the Nagoya Protocol on Access to Genetic Resources and the Fair and Equitable Sharing of Benefits Arising from their Utilisation to the Convention on Biological Diversity aim to prevent this. Scientists and researchers at the Royal Botanic Gardens, Kew, are developing ways to work ethically and fairly with local communities and their plants, and actively seeking to address their legacies and futures. However, the uses of plants such as

Cinchona, are complicated to unpick. There is still a long way to go on allowing fair and sustainable access to and conservation of plants at the same time as supporting local people and their knowledge and preventing their exploitation. Quinine and *Cinchona* bark are still sourced far from its original roots in the Andean cloud forests, and profits are still diverted away from the local peoples who originally lived under the tree. Nowadays *Cinchona* trees are mainly grown commercially in plantations in Indonesia and the Democratic Republic of Congo. Although malaria is now treated with other drugs, including artemisinin from sweet wormwood (*Artemisia annua*), quinine is still found in hospitals, used in cases of malarial resistance and for other diseases such as babesiosis. It is also produced for popular beverages like tonic water.

In South America, *Cinchona* is under threat from climate change and habitat loss, including competition from coca/cocaine (*Erythroxylum* spp.) plantations. Though it is not listed as endangered on the IUCN Red List, this may be down to poor understanding of its status, as well as the knowledge that its mass cultivation means it won't become extinct. Records reveal the rarity of the tree nowadays compared to the seventeenth century, when it was first harvested on a mass scale – a state of affairs confirmed by botanists searching for it in the wild today.

Thankfully, *Cinchona*'s continued popularity means that in South America some are seeking to raise awareness about its history and reclaim a chapter in the plant's history. At least one conservation project has been established by local environmentalists and researchers: Semilla Bendita is an ethnobotanical garden in San Ramón, Peru, which hosts local educational projects and grew 2,021 *Cinchona* trees for the two-hundredth anniversary of Peruvian independence in 2021. Another project called *Rescatando el 'Árbol de la Quina' en La Cascarilla* in Jaén, Peru, has brought multidisciplinary researchers together in symposiums and regular meetings, the better to understand the *quina* tree and its chemistry, genetics, conservation and cultivation, from non-Eurocentric perspectives. Many of these projects focus on conservation and educating local people. At the same time, new jobs are created to provide seeds for commercial use.

Plants can be powerful chemical factories, but they don't perform this function in isolation. All life is interconnected, and interacts in complex ways we cannot predict. As we move forward into a post-pandemic world, plants will remain important sources of new compounds and cures, but we must be aware of how they can be used sustainably and, most importantly, fairly, so that, while everyone can benefit, those who hold the knowledge and biological origins of the plant, and the communities around them, are supported without exploitation, and their practices and ways of life respected.

NATALY ALLASI CANALES ADDS:

In Quechua culture, we are proud when we listen to stories of the *quina* tree [the Quechua name for it, and the origin of the word quinine] saving peoples' lives beyond our Andean communities. However, we are also deeply concerned about the lack of information on the conservation status of this iconic tree, since we know it has been over-exploited for centuries. The plants of our nations have to be protected, so their ecosystems are not threatened, they are not used as tools of control, and the people most closely associated with them are not subjugated by excessive economic and political forces.

As a scientist working on the *Cinchona* tree, which is depicted at the centre of our national flag, I was surprised how much work needs to be done to understand the different processes this genus has gone through – naturally, and as a result of human activity.

In the last few years, research groups and independent researchers from Andean countries and backgrounds have begun to collaborate, with the aim of focusing attention on the conservation, management and restoration of the *Cinchona* forests in South America.

During our research on the *quina* tree, I have encountered barks that were collected from the Andes in earlier times and are now found in botanic gardens, museums and herbaria around the world – so far away from home. For me, imagining the multitudinous journeys the bark collections I study have been on was an overwhelming experience. As I handled them, my inner scientist was amazed by the richness of the data we can access with state-of-the-art technology, such as the genomic and chemical analyses I myself carry out. There is so much still to understand, for example, about the genetic origin of *Cinchona*, and by extension the potential chemicals it can yield for future drug discoveries.

As an Andean, indeed, I feel that by restoring this tree to its rightful place as a key part of our history, I could myself bring about a little bit of justice. And it's this that has kept me going during all those moments when experiments fail: the thought of what this tree means to my people, and that one can only guess at the fascinating results we might get ... There is no room for failure.

However, as a descendant of the Quechua peoples, I also feel a deep melancholy for a world which is being ruthlessly depleted and yet still resists complete destruction. Hundreds of years later, it is we, once again, who are attempting to rescue our tree from extinction – may Pachamama help us. Still our languages, costumes, traditions and bitter barks thrive in a world that persists in forgetting humanity's strong connection with our environment. Still our plants provide us with endless resources, and we shall do all we can to help them thrive.

8

HOW TO STUDY THE MOSSES

Jessica J. Lee

The great advantage in studying the mosses is that the material is at hand, in winter as well as in summer, in the city as in the country, that it can be used when dry and old just as well as when fresh, and may be sadly ill-treated and yet when soaked out will be as good as when fresh.

Elizabeth Gertrude Knight Britton, 'How to Study the Mosses: Part I'

1 COLLECTING

IN A CLUMP BENEATH my monitor sits a mass of faded green. I often forget that it's there, a tiny sign of life against the metal and wood of my desk, rescued from the pavement during a winter walk last year.

In December, clumps of springy turf moss started appearing on the footbridge between the park and the marshes, nosed up from the kerbsides by the boots of walkers. I couldn't just abandon it there on the concrete, high above the thoroughfare and its traffic of white vans. I lifted a clump and pressed it into my pocket, warm and safe.

I'm not entirely sure when my obsession with mosses began. But I remember a day in the early 2000s, when my mother and I stood in front of a brick wall simply noticing the green that grew upon it. My mother is a keen gardener, and a lover of flamboyant plants, so to me her love for mosses seemed somewhat odd. I expected her to notice orchids or azaleas, not the smear of green on a roadside wall. But she said two things that I have not forgotten: the first was that mosses were amazing for growing practically anywhere. The second was that they reminded her of home.

2 PRESERVING

What does it mean to love something that is liminal? That falls out of sight so easily, yet is integral to nearly any landscape you can think of? Moss exists at the border between worlds: the evolutionary bridge between algae and higher land plants, mosses were the first venturers onto land. This strange existence means they aren't quite like other plants. They don't reproduce by flowering like most plants we know well, but they still occur on every continent of our planet, reproducing by spores in a manner not unlike mushrooms. Bryophytes – the mosses, liverworts, and hornworts – don't even have roots.

Clutching to bits of rock and wood using filament-like rhizoids, tiny hooks that clasp the surface on which they grow, mosses remain a relatively portable thing: dislodged from their original place, they can thrive again almost anywhere suitable, given shade and water. Dehydrated to the brink, they can grow green again with rain. Plunged into extreme heat or inhumane cold, mosses wait quietly and then, when conditions are right, return to the act of living.

It seems like a crass overreach to compare plants and people, but the portability of moss strikes me as more human than vegetal. Three generations of my family had migrated – from China, Taiwan, Wales and Canada – and all had established themselves in a new place, seeking shelter, sustenance and light. I once carried a cluster of swan's neck thyme moss in my pocket for months, only to revive it with a bit of rainwater and a new home in a potted plant in the courtyard outside my apartment. At the time, it seemed to me more resilient than a plant should be. But mosses, unlike people, have little to no agency in where they wind up. They are movable but cannot move themselves, and perhaps that is their vulnerability.

3 STUDYING

If mosses are perceived as peripheral to our world, then the tales of women who study them are even more so. Histories of women operating at the fringes of nineteenth-century science are many: from the algologist Margaret Gatty to the plant cyanotype artist Anna Atkins. Most often, it was overlooked plants like mosses, ferns and seaweeds that women were able to stake some claim in. This was, in part, because these plants were deemed more suitable for polite women: often reproducing by spores, they were seen as *less racy* than flowers that produced by sexual means. But, still, most women scientists of the era were forced into a kind of informality. Agnes Fry, who illustrated her father's 1892 book *British Mosses* and in 1911 co-authored the subsequent *Liverworts*, warrants hardly a mention in histories of bryology, and only in reference to her father's accomplishments. Fry, Gatty, Atkins and others were tolerated as amateurs, the women given quiet consent to occupy a space at the edges of male science.

In an eight-part magazine series from 1894 entitled 'How to Study the Mosses', Elizabeth Gertrude Knight Britton recounted the basics of bryology, microscopy and collecting specimens of moss. Britton was among the most prolific bryologists of her time, and was a crucial figure in establishing the study of mosses in North America. But what strikes me about her career is that for the most part, despite her extensive research and knowledge, she was never able to professionalise her work as a man might have. She was the 'unofficial' head of bryology at the Columbia College (now University)

Herbarium, despite over the course of her lifetime publishing over 300 papers about ferns, mosses and wildflowers.

One obituary of Britton described her as full of 'feminine impulsiveness', wavering between generosity and criticism. Generous, it seems, to a fault: the author of the obituary points out that when he published his own guide to mosses, Britton told him that she had also once planned to write such a guide, 'but she forgave [him] sufficiently to assist', setting aside her own professional achievement for that of another. On a visit to the Linnaean Society, she was asked to work upstairs, not on the main floor, because she was a woman. Like the mosses she studied, Britton's research fell to the periphery of the milieu in which she found herself.

So how, then, might we study the mosses? Britton suggests staying close to the samples themselves: collecting, dissecting, preserving and observing. I would add one further provision: to understand mosses – not just bryologically-speaking, but culturally, symbolically, emotionally – we must likewise understand the history of their study. If they are the bridge that brought life to land, we must ask, why do we see mosses as peripheral?

4 TEXTBOOKS

In the autumn of 2014, after a day spent in a Brandenburg forest counting more mosses than I had ever seen before,

I struck upon the idea that I might like to write a cultural history of mosses. I understood them so little, and wanted more than anything to devote my time and being to a world so small it could go unnoticed. As if in doing so I might dissolve my own sense of self, my own scale in a world far vaster than me.

The first thing I did, of course, was google to see if someone had already written one. No less a luminary than Robin Wall Kimmerer already had: *Gathering Moss: A Natural and Cultural History of Mosses* was published in 2003. My momentary disappointment at not having happened upon the idea first dissipated quickly. I ordered the book and spent all of that November in its pages, savouring more slowly than I ever had before. Because in Kimmerer I found not simply a science of mosses – their life cycles, structures, needs and adaptations laid out in careful prose – but an ecology of their emotional resonances, too. The pleasure in feeling wind blow between the moss-covered boulders in a field survey plot. The weight of a basket on the arm as the bryologist collected samples. The sensation on the tongue of many thousands of words for moss – from Indigenous ways of knowing to the taxonomic. Mosses, here, did not recede into insignificance owing to diminutive size. Mosses opened worlds for the knowing.

5 MICROSCOPIC CHARACTERS

The boundary layer, writes Kimmerer, is 'the meeting ground between air and land'. If you lie close to the ground, the wind deadens and moisture surrounds. This is the trick that makes moss possible, that keeps it thriving in winter, when everything else shatters in ice. The microclimate of moss traps moisture – the most precious thing to a plant whose cell walls might be just one cell thick.

In 2015, I took a field survey course during which we spent a day on the ground, cataloguing everything within a one-metre-plot of Richmond Park. I wore rubber boots and a raincoat, my 10x hand lens strung around my neck on a loop of butcher's twine. For two weeks we'd used our lenses to identify ferns, grasses and flowers dried and pressed on herbarium sheets. I wanted nothing more than to feel damp soil beneath my fingers, to find between the grasses a world even smaller than my lens could properly magnify.

I should say, I loved the course. I learned more in two weeks of hands-on study with plant samples than I could have from months studying pictures and plant keys in flora. But no one brought up mosses. There was no bryologist on the team. Mosses are not vascular plants, and for that – and their microscopic size – they are neglected. Mosses are simply not viewed as a priority.

But I believed there was a lesson in learning to look closely.

On my knees, in the middle of Richmond Park, I threaded my fingers through the hook of my lens and counted: seven types of grass, three types of moss. As the instructor leaned in and named each type of grass – a feat, given how small the flowering bodies of grasses are – I lingered over the mosses. I knew that without a microscope the chances of really identifying and understanding them were slim. But still I wanted to learn how to look, how to look and not even name. Was that simply appreciation? I pressed my camera to my hand lens and snapped a photograph.

6 CLASSIFICATION

The final directive in Part I of Britton's 'How to Study the Mosses' is to classify them. But barring becoming bryologists ourselves, how might we go about this act of naming? Mosses are small and difficult to parse from one another. In years of learning, I've memorised only a few: *Mnium hornum, Rhytidiadelphus squarrosus, Dicranum scoparium, Polytrichum commune.* They are the mosses I see often, the ones I know by eye and touch, from long hours spent watching the forests. But I don't think of them by their names all that often. Instead, I think of their shapes: stellate or forked, long sporophytes reaching into their upper atmospheres. I think of their shades – russets and golds amidst greens – and how much they change with the weather, with rain.

Mosses are pioneering species. When a landscape is denuded by wildfires or rockfall, it is mosses that return to the soil first. Venturing onto the scene, they call so many other species – invertebrates, birds, vascular plants – in their wake. They do this in cities, too.

When my mother told me that mosses reminded her of home, I didn't immediately understand. But then I travelled to her home – to Taiwan – and watched for mosses as they made their lives in every small space. They grew on tiles, on windowsills, on the barks of city and mountain trees. They grew on scree where mountains had been scuppered by landslides. So I learned in watching for them how I might classify mosses: as resilience, as patience, as home.

NOTES

92 '... they were seen as *less racy* ...': Ann B. Shteir, *Cultivating Women, Cultivating Science: Flora's Daughters and Botany in England, 1760–1860* (Baltimore: Johns Hopkins University Press, 1996).

'... only in reference to her father's accomplishments.': 'The Right Hon. Sir Edward Fry, GCB, FRS', *Nature* 102, pp. 169–170 (1918).

'In an eight-part magazine series ...': Elizabeth G. Britton, 'How to Study the Mosses, I', *Observer* 5, no. 3, pp. 82–86 (1894).

93 'One obituary of Britton . . .': A. J. Grout, 'Elizabeth Gertrude (Knight) Britton', *Bryologist* 38, no. 1 (1935), p. 1; www.jstor.org/stable/3239293.

'. . . "but she forgave [him] sufficiently to assist" . . .': Ibid., p. 3.

95 '. . . "the meeting ground between air and land" . . .': Robin Wall Kimmerer, *Gathering Moss* (Corvallis: Oregon State University Press, 2003), p. 15.

9

A PLANET WITHOUT FLOWERS

Sumana Roy

'PEOPLE FROM A PLANET without flowers would think we must be mad with joy the whole time to have such things about us.' Iris Murdoch's words about flowers inevitably lead me down a side alley. After the inescapable thought about her being named after a flower that returns every single time I run into this quotation, I think of 'people from a planet without flowers' and what they would make of the relationship between flowers and love.

In the India that raised me, one before satellite television and skin, actors rarely kissed on screen. Two flowers – usually roses – did. They pecked each other. The actress became pregnant soon after. Even when it was just the red or white roses brushing each other, many of us looked away from the screen. We were with our parents, after all. Sometimes we would walk away right at that moment to get a glass of water. Flowers were central to the reproductive life of plants, but it wasn't that theoretical knowledge that was responsible for the charged energy of the living room. There must have been something about the flowers.

In the mornings, as grandmothers and flower thieves set out to gather flowers from the neighbourhood for the worship of their Hindu gods, the sexually charged energy of the previous evening was gone. In baskets and paper packets, and often in the anchal of their sarees, the flowers, usually white – the garden jasmine and beli and hibiscus, if they were lucky – looked timid, even domesticated. They resembled broiler chicken to me – waiting for their turn to be sacrificed.

Even for someone like me, largely indifferent to and even a tiny bit annoyed by the glamour of flowers, there was seduction. I was aware of my role as a pollinator, but it was still hard to ignore the question – why, why this ferocious beauty of flowers? Science has its answers, but what about lovers? Why had flowers come to be part of the cosmology of love?

I decided to go to Kama Sutra, perhaps the human world's most well-known literature on love and its many experiments, to find whether it had anything to say about the relation between flowers and romantic love. Inside my copy – hidden behind other books to avoid suspicion of my proclivities – was a pressed flower. The neelkantha, its blueness darkened by time and the pressure of leaves (the Bangla word for 'leaf' and 'page' of a book is the same: paata). How this flower – blue pea, butterfly pea, Darwin pea to the world – has ended up in my copy of what is mostly read as a sex manual I cannot remember. Gossip about the flower returns to me, not completely unrelated to why it is one of the god Shiva's favourite flowers: how its form is similar to the female genitalia. I suppose one could imagine this – and its obverse – for many flowers, but the visual analogy had caught on for some reason.

For a moment I am distracted. I try to remember other flowers and their histories that I have met elsewhere, flowers with which the history of human emotions is connected in minds like mine. The flower of the Ashoka tree – a-shoka, without sorrow, the sorrow-less tree – that one cannot miss on gates in Hindu, Jain and Buddhist architecture. There's the lavender, of course, which I haven't seen on fields, but which lives on pillowcases, coaxing rest and sleep out of my eyelids. Red flowers, now commended by scientists as contributing to the health of the immune system, energised me as I cycled past the string of florists when I was a schoolgirl. I remember some of the exhaustion of the school day draining away when I saw red roses on the shop sills, gathered into tight bunches, a bit like the ponytail my mother had forced my hair into in the morning. There were also the garlands of red hibiscus that hung from nails on the shop ceilings, waiting for devotees and worshippers of the goddess Kali at temples nearby.

Beside them, for most of the year, though they are seasonal flowers, were marigold garlands in two shades of yellow. On a dark and dull wintry day, one could mistake them for fairy lights from a distance – they illumined the shop and the day, their effect staying on long after I had biked past them. Once, in a bus to college, I heard two young lovers fighting – a lover's quarrel. Something insignificant had happened and then taken a cyclonic shape, as such situations between lovers often do, for love derives its intensity from the voltage of the present moment. Every time the bus slowed at a traffic signal, it seemed to me that one of them would get off, such was the centrifugality of the moment. And then the traffic crossing – called Venus More, after an old hotel there – came, and suddenly the girl said, 'When I see these flowers, I want to get married.'

I looked out of the bus window. Spring had arrived unexpectedly at the florist's. Roses and tuberoses were now on the bodies of cars – for in this country it is not only the bride and groom who dress up patiently for the wedding; the car that carries them must as well. As the bus slowed down in the traffic, the young couple rushed out of it. I kept looking for them in the crowd. From far away I saw them jaywalking through the busy road to what looked to me like a mobile garden of flowers. I don't know whether they got married right then. If they did, it wasn't only love that made them do so. It was the flowers, the flowers.

Just as it was the forest and its residents that was responsible for the irrepressible romantic attraction between a king and the adopted daughter of a sage in the poet Kalidasa's play *Abhigyan Shakuntalam*. Shakuntala, the beautiful young woman, lives in an ashrama. The word derives from shrama, meaning labour, but it comes to mean 'religious exertion': not penance, but a way of living. The ashrama was usually in a forest, away from the distractions of a materialist life. Even though Kalidasa is writing this in the fourth to fifth century AD, there is distantiation in space and time between the present moment he inhabits and the world he is showing to his audience. The paradisal idea of the forest as a counterpoint to the urbanity of the capital city of the kingdom where Dushyant, the king, lives, is already an anachronism when Kalidasa writes about it more than 1,500 years ago.

Right from the moment when we first meet the deer, the chasing of which eventually leads Dushyant to run into Shakuntala and her evident beauty, which nearly stupefies him, we are given a world that is experienced – or perhaps can be experienced – only analogously through plant life. Actually, I think it starts even earlier, in the prologue itself, when the 'actress' draws the 'director's' attention, for no real reason, to the 'sirisa blossoms ... crested with delicate fila-ments'. When we meet Shakuntala, one of her friends calls her 'as delicate as a newly-opened jasmine'; she responds soon after with a complaint – that her 'bark-garment' has been tied too tight. The king, someone from the city, sees her as a plant – 'though inlaid in duckweed the lotus glows,' 'her lower lip has the rich sheen of young shoots, her arms the very grace of tender twining stems; her limbs enchanting as a lovely flower.' This continues throughout the play, and one is led to wonder whether much of Shakuntala's unease and suffering, at the king's inability to recognise the woman he had loved and married, is because of her being uprooted from her forest life, and her consequent withering from neglect from being transplanted into a new world. This metaphorical fluidity that compelled humans to see their own species as composed or made of plant life gave us a world – and a world view – where neither plants and animals nor the elements were seen as antagonists. They were all co-sharers of the earth.

'Kama,' Vatsayana's translator A. N. D. Haksar tells us, 'is the mind's inclination towards objects which a person's senses of hearing, touch, sight, taste and smell find congenial ... a delightful, creative feeling pervaded by sensual pleasure and derived in particular from the sense of touch.' Kama Sutra (kama: erotic love; sutra: theory) mentions sixty-four subsidiary subjects that one should know to become a skilled lover. It is an interesting list – interesting because one pauses to wonder how being able to make parrots and mynah birds talk can make one an efficient lover. But there are also the ways in which the plant world is invoked and invited to be a participant in and indulger of erotic love:

making decorative patterns with rice grains and flowers; arranging flowers; making garlands; cooking recipes using uncommon vegetables; making toys with flowers.

Luck in love can come about
with just a knowledge of these arts.

There are prescriptions about habitat as well, and it is striking that in them too the proximity with plant life is emphasised: the house should be near a water body and should definitely have an orchard; near the head of the bed should be, besides beeswax and bottles of perfume, lemon peel, garlands, betel leaf, a garland of amaranth leaves; a swing in the orchard and a bench with flowers spread on it. Then there are the 'daily routines': the man must apply scented paste, most possibly sandalwood, wear a garland or two, chew betel leaf to freshen his mouth.

Book Two of Kama Sutra is about the various kinds of union possible between humans. We learn that sexual union has sixty-four elements, and that among the many preludes to sex there are some that import the optic of plant life. In the 'twining creeper', as it is called, the woman twines herself around the man like a vine wraps itself around the sal tree, and pulls his face towards her to kiss him. She leans against him and looks at him with affection and longing. The metaphorical borrowing from the plant world to express the union of human bodies is a surprise, but this is not a lonely analogy. There are others that follow. There's 'climbing a tree', where the woman holds and clasps on to the man while both are in the standing position; she sighs and she pauses, she breathes heavily, sometimes stopping to catch her breath, for all she is trying to do is to kiss the man, for which she has to climb on his body as one climbs a tree.

What is impossible to not notice is the easy interchangeability of human and plant, both species borrowing from each other in the human imagination, the fluidity between beings conceived so naturally that I cannot help wondering how far the distance between both over the last two millennia might be responsible for the state of the world today. In the same chapter in Kama Sutra is an explanation of another kind of union: the 'sesame and rice', where the man and woman lie in bed, their limbs wrapped tightly around each other.

An entire chapter is devoted to 'scratching', including it as necessary to making love. Nail marks are categorised: most are named after wildlife – the 'tiger's claw' and the 'peacock's foot' and the 'leaping hare' – but, right at the very end, unexpectedly, there's also the 'lotus leaf'. It is one that the man makes on the woman's hips or breasts – several marks close together. This has symbolic value: she is to see the lotus leaf on her body and remember him, for he is now away, travelling. Besides scratching, there is biting – not only on the body, but also on flowers and leaves. It is a code – bites on floral ornaments for the head and ears, on betel and bay leaves for a man showing interest in a woman – his advances, as it were. These marks, on bodies and flowers and leaves, are like urgent telegrams of passion.

The lotus occurs elsewhere as well, in the chapter on intercourse – the woman crosses her legs and sits on the man as if he were water and she a lotus. If the woman puts her foot on the man's shoulder and repeats it with the other foot a few times, it is called 'splitting the bamboo', for that is how a sharp instrument cuts through a bamboo. In the section on 'hitting' during making love, the sound made by the woman is compared to a bamboo being split and her crying to a berry falling on water. It is as if Vatsayana is constantly looking sideways, turning to look at the trees around him to explain the metaphorical and physical union of two humans in passion. This becomes all the more moving because the passion of the plant world is a secret kept from us.

The matter-of-factness of comparing a kind of oral sex with sucking a mango, of girls to flowers who should be treated delicately, of turning the betel leaf into a vehicle and host of passion, the chewed leaf being exchanged between the two lovers repeatedly during kissing, or the man making figures and designs by chewing leaves to indicate his desire to the woman, the kusha grass and fire around which their marriage is solemnised – this world does not seem anachronistic as much as idealistic. It is, after all, an ideal world and an ideal way of lovemaking that Kama Sutra wants to hold before us, a manual that might bring more beauty and passion to our lives. The resistance today is not so much within ourselves, polished in different ways by ideologies and ways of living and loving, as inherent in the disconnect from the plant world. The instructions would seem incomprehensible without a basic familiarity in living beside orchards and forests, even within them.

In Book Four, called 'The Wife', we see what is now a world accessible to us only through cultural memory and the imagination. One of the responsibilities of being a wife is to maintain a well-weeded garden. And to include in it at least the following: sugar cane, herbs, greens, cumin, fennel, aniseed, cinnamon, gooseberry, magnolia, hibiscus, jasmine, flowering plants, various kinds of grass, radish, potato, pumpkin, cucumber, brinjal, gourds, beans, onion, garlic, among other species. The woman, who is expected to have the skills of a farmer, also – lest we forget – wears flowers in her hair.

Vatsayana also mentions the use of vegetables – roots and fruits of 'appropriate' shape – to act as a dildo or to cast a spell by being cooked in a certain way. Roots and flowers and leaves are boiled and burned to bewitch a woman or enhance a man's virility, things mixed together to create potions, all of it pointing to a culture and temperament of experimentation that believed in everything being worthy of attention and seriousness and concomitant humour. These practices make us relatives to our ancestors of centuries ago – for among the many things that connect us is our awareness of sex as something fundamentally comic and beautiful.

As we move to a world without bees, which would, consequently, leave us with a world without flowers, I think about the last days of love on this planet.

ECO REVENGE

Susie Orbach

CLOTHES WERE ONCE made from plants. Or mostly so. Silk, cotton, wool, hemp, flax. Today, polyamides made from crude oil create nylon. Lycra is made from polyurethane. The clothing industry, from production to the labour processes, to the distribution of clothing, to the recycling of unwanted clothing, has a huge economic, ecological, cultural and personal impact.

Clothing is not simply a covering of the body. It is a statement of belonging, of aspiration, of signalling. Where once clothing was a class, ethnic and geographic marker and thus highly differentiated, late capitalism and visual culture has collapsed variety in clothes. The hoodie, high heels and sneakers, have become ubiquitous worldwide, as has the suit, jeans, the cocktail dress and so on. We are encouraged to think we need them all. Images of taste makers saturate our visual field, subtly creating a sense that our personal duplication of their style is a requirement for cultural acceptance.

Behind the imagery are huge industries which pollute our minds and pollute the planet. Their labour practices are exploitative, whether in Sicily, Bradford or Bangladesh. While recent stories have condemned the pittance wages and piecework rates that companies like Boohoo pay, leaving the impression that this is a feature of fast fashion, at the high end the picture is not so very different. Southern Italian garment workers sew by hand for upmarket brands, their daily wages equivalent to two hours of the UK minimum hourly rate.

It may startle some to know that two of the wealthiest people in Europe are in fashion. Amancio Ortega, the head of Inditex, which owns the brand Zara, is reputed to be worth $77.7 billion (June 2021) while François-Henri Pinault ($42.1 billion in March 2021), who besides owning Gucci, Stella McCartney, Boucheron, Alexander McQueen, Bottega Veneta, Yves St Laurent and many other brands, also owns the auction house Christie's. These are not trivial industries. These are not small industries. These are nimble industries that know how to source material, pattern-cut, manufacture, sew, label and distribute goods within a very short time frame.

Shein, the new privately owned Chinese company, has refined this process. From production to appearing on the website for sale takes only a week. We are used to car parts and agricultural produce being 'just in time'. Doubtless Shein and its rivals will move to bespoke, or perhaps I should say made-to-measure, and future apps will make it easy for us to order a garment for next Saturday. Just-in-time clothing exposes our relation to clothing. It is deemed essential and time-critical. The logistics which apply to the delivery of shampoo, eggs and car parts now apply to clothing. Fashion has a short shelf life in which the wearing of a party frock once – previously the preserve of the wealthy – has now been democratised. Young women and men post pictures of themselves in an outfit which thereby render that outfit redundant and unwearable. What can unwittingly become effaced are specific cultural markers. That's not to say that human ingenuity and resistance can be stifled. Subgroups an individual may identify with or aspire to belong to code themselves and subvert the proposed uniformity. Nevertheless, along with cosmetic surgery, from jaw shaving to eyelid insertion, nose reduction, breast augmentation or reduction and height enhancement, the body itself is now being constructed. Late capitalism in this global moment takes the body as a site of production and a site of revenue. We labour on our bodies, and in doing so we provide the profit for nefarious industries, of which clothing is one.

When we think about production, we are reminded of the Rana Plaza fire in 2013 which killed 1,134 women garment workers. We recall the conditions of work at BooHoo's overcrowded warehouse in Leicester where, during the first lockdown, there had been no Covid-safe measures in place. Leicester became one of those towns with a high percentage of Covid cases. We can go back to the Triangle Shirtwaist Factory in 1911 in New York City, which again had dangerous overcrowding, in this case resulting in a notable and murderous fire. These standout moments aren't particular. They are part of a pattern of exploitation of garment workers which yields fabulous profits for owners. The *Guardian* reported that in the twelve months to 28 February 2021 Boohoo's sales climbed by 41 per cent to £1.7 billion, up from £1.2 billion a year earlier. Its adjusted pre-tax profit climbed by 37 per cent to a better-than-expected £174 million. Meanwhile, machinists and warehouse workers are alleged to be on less than half the minimum wage, at around £3 an hour.

So profits of an extra half a billion pounds in a pandemic year, when no one was going anywhere, should give us pause. Clothing production produces appalling social, economic and planetary despoiling.

If we start with production processes, we know that from the water that is diverted to grow cotton, to the dyes that make their way into the water supply, to the shedding of

Lycra fibres which enter the sea and contaminate and choke many fish, we pollute. Add to this the cost of distribution by land, sea and air and we have a cocktail of toxicity. Consider that we are re-ingesting microplastics into our guts which end up infiltrating the breast milk fed to the next generation. It might be called eco-revenge, only this kind of revenge is not consciously targeted: it seeps into our water supplies, the sea, what we eat and our lungs. It is of a piece with climate change, uncommonly extreme temperatures, unexpected winds and intense rains.

Lest we get succour from thoughts of virtuous recycling, we had best reflect on the damaging effects of our charitable giving being shipped abroad to African countries such as Burundi, Kenya, Rwanda, South Sudan, Tanzania and Uganda. Our second-hand clothes are unwelcome, we learn, as they destroy Indigenous clothing industries that have developed modes of production which are more in harmony with local conditions, and which don't rely on slave-like labour to produce cotton. We can forget that mill owners were slave traders, importing and exporting indentured men and women to work the cotton milled in Britain and grown in the American South.

The underbelly of the clothing industry belies the patina of glamour. Culturally, our museums may display the art of the dress and elevate it as high culture. Yes, clothing is expressive and imaginative. But as we have seen, even the high end of the industry cannot serve as a fig leaf for environmental and social destructiveness. High-end clothing as sold on Old Bond Street for women contains considerable amounts of plastics at the same time as it is demarcating the class and wealth positions of its wearers. Clothing and fashion – not just fast fashion – demand scrutiny for environmentalists for the multiple damages they may unwittingly inflict.

Over the last few years, some designers have started to use sustainable fabrics. They have also started designing for the variety of bodies that actually exist, as opposed to the ones that are photographed and curated for aspiration and exclusion. Some talk of slow fashion, but of course in economic terms that is a difficulty for the industry. It has become accustomed to fast turnover on the one hand, and low-paid workers – cutting, sewing, working in magazines, interning at fashion houses – and non-biodegradable materials on the other. Ellen MacArthur's foundation has been arguing in her work on the Circular Economy for a rethink of the industry from the production and disposal point of view. Tansy Hoskins' books *Stitched Up* and *Foot Work* deserve a wider readership for their analysis of the fashion industry. Meanwhile the activist group AnyBody has since its inception, when it demonstrated at London Fashion Week in 2001, been arguing for sustainable clothes on sustainable bodies.

Consciousness is changing, but this is a big and powerful industry, which easily flatters the pretensions of the design world. That world needs puncturing. The pleasures it brings are not straightforward; they are entwined with social, psychological, economic and ecological damage.

NOTES

120   '... Ellen MacArthur's foundation has been arguing...': www.
ellenmacarthurfoundation.org/our-work/activities/make-
fashion-circular.
'Meanwhile the activist group AnyBody...': www.any-body.org.

11

NATURE AS HEALTH

Araceli Camargo

HEALING IS SACRED. Healing is the lifelong process of approaching wholeness. We heal with, and for, Nature. Nature, with a capital N: a being, our kin, a spirit, an energy. We, too, are Nature.

As our hands work through soil, when mending a plant, we inhale millions of microbes that provide vital support to our gut ecosystem, ensuring the healthy function of our various bodily systems. As we take the time to heal the plant, the plant heals us. When we find our way through the knotted and uneven ground of a lush forest it helps us create supple, resilient and flexible minds, muscles and skeletons, keeping us in a healthy active state. The more we do this, the more we create the capacity to keep doing it. We heal *for* Nature. We heal to sustain our ability to interact with Nature, to know this plural being, to care for them, creating a lifelong healing relationship.

We are kin. All living beings which inhabit this earth are contained within nature, from rivers to plankton to people. Nature holds multiple energies and lives. There is no divide or hierarchy between us: every living being has a contribution to give, has worth, and the right to exist with dignity and health.

Oxygen binds our kinship. Time and space created an opportunity for Oxygen to enter our Planet, survive and thrive through humble, single-celled organisms called cyanobacteria. This miraculous phenomenon gave the foundation for all life. Every single being needs and contains Oxygen, making it one of the most omnipresent elements on earth. Oxygen is in our Waters, Land, Air, and in our blood, creating an unbreakable bond between us all – an eternal kinship with all living things. Oxygen promises continual, abundant and healthy life.

Health is a process. It provides our biological systems the opportunity to create a stable state after experiencing trauma or stress in our lifetime.

When our habitats are healthy, we are healthy. Health does not happen in a vacuum, or within the confines of our bodies. We heal with Nature. In a richly biodiverse environment, such as a forest, our respiratory tract allows us to ingest microbes from vegetation, fungi and soil. These microbes are essential to the health and function of our internal microbiome found in our gut, and are known as *supportive capacities*. As biological sciences evolve, there is an increasing body of work linking the function of our gut's microbiome to complex diseases such as depression, anxiety, obesity, dementia and diabetes. Exposure to biodiversity supports our health in preventing disease, but also helps lessen the burden of disease. From Nature we can receive vital nutrients and medicine that help us heal.

Nature can also provide *restorative capacities*: outdoor habitats like forests, rivers and mountains can rehabilitate us from the burdens of the lived experience. From a cognitive perspective, these habitats can create the experience of fascination, which is linked to effortless attention, reducing the demand on our cognitive resources, allowing our mind to restore. This is in contrast to the sustained attention required to successfully navigate a busy street, where there is a high demand on our cognitive capacities, owing to the varied and unexpected events, stimuli and activities our minds have to make decisions on and decipher. And from a physiological perspective, when we are in richly biodiverse environments our body's stress response is better regulated. This biologically adaptive response allows us to adapt to acute and unexpected changes to our environment. In our modern world, where stressors can be constant and our stress response is in high demand, this can cause wear and tear on many of our biological systems. Nature's restoring qualities can be important for a lifetime of health and healing.

And thirdly, Nature enables us to connect. *Connectedness* to Nature is our ability to mentally, spiritually and intellectually bond with the natural world. The feeling of belonging is very important to all beings. Our sense of belonging is not limited to human contact, we also need contact with the wider natural world. All beings on this Earth need to be connected and bonded in a kinship as it secures our collective survival. Loneliness, which can come from a lack of connectedness, is part of the disease pathology for depression, anxiety, obesity and PTSD. Nature, and in particular a richly biodiverse environment, helps us heal, helps us connect, helps us feel whole.

Sadly, the biodiversity of our Planet is becoming more and more depleted and sick. Robust, healthy trees are becoming diseased from air, soil and water pollution. Precious ecosystems like coral reefs are dying, and all our life-giving kin – Water, Land and Air – have become polluted. There is no place on earth free from plastic; even the snow in the Arctic contains microplastics.

This is driven by extractive economies – economies dependent on the commodification and industrialisation of natural resources, including human labour. They are taking away our time and space to build kinship with Nature, taking away opportunities for healing. In such an economy, society has one concern: profit, by any means necessary. Profit drives decisions. People are paid non-living wages in order to increase profits. But this can have a knock-on effect on health. Non-living wages force people into dangerous housing. Social housing is often built next to industrial waste sites or busy highways, because it is cheaper – but it exposes people to environmental pollutants. Non-living wages force people to live with psychosocial stressors such as food, housing and job insecurity, and various types and scales of trauma that are part of the poverty landscape.

This is *biological inequity*: the systemically driven, uneven distribution of biological stress in a population. Those experiencing poverty and environmental pollutants are constantly forced to engage their stress response, leading to the dysregulations of our cardiovascular, metabolic, endocrine, immune and digestive systems. Research is now beginning to show us how these dysregulations play a key role in disease pathology.

Health is, therefore, an ecological phenomenon. Health is intrinsically linked to the places we inhabit. If we want human health to flourish, we must start by allowing our Planet to heal.

The phenomenon of poor health we are now experiencing was not inevitable: it was deliberate and orchestrated. To understand why, we have to travel back to the rise of feudalism in Europe in the fourth and fifth centuries, because with it came the degradation of the land as the aristocracy sought to make the land profitable. They exhausted their labour force, forcing them into poor living conditions, while at the same time they overworked the soil with no regard for its health and restoration. It was a disastrous combination that left both people and land vulnerable to disease.

Current capitalist structures create a very similar poor health landscape, which is why we are seeing very similar outcomes. Societies built on extractive economies that have enacted colonialism and continual imperialism have disconnected us from the Land. Indigenous peoples of Turtle Island – my people – have been ripped from our Ancestral Lands, causing some of us to lose our knowledges and kinship with the Land, knowledges that are vital to traditional methods of healing. However, it is only an interruption – we are always finding ways to reconnect: we do so every time we tend to plants, introduce children to the notion of biodiversity, protect Water, Land and Air, and think of our Earth Mother in a kind and loving manner.

Nature was never meant to be for profit. Nature was meant for healing.

It is crucial to acknowledge that extractive economies have not gone unchallenged. Throughout history many Indigenous peoples all over the world have resisted; a resistance built on knowledge and imagination that prioritise kinship over supremacy. This is evident in the fact that most of the world's biodiversity exists in territories sustained by Indigenous peoples. We can, and must, move away from the governing supremacy in knowledge that has permeated the cognitive architecture of so many societies.

We need to heal. We need to imagine a future of healing.

And to do so, we need different ideas, different imaginations, different knowledges.

Imagination is important. It allows our minds to build new cognitive architecture, from which to form socio-cultural decisions. But we cannot keep upholding knowledge supremacy: it has created a homogenous, discriminating and dominating knowledge infrastructure that decides which knowledges are valued, listened to and acknowledged – and which are not. We need to embrace *imaginations* in order to heal – from all peoples, from all living beings.

We need Indigenous futures in our imaginations. Indigeneity, as a word, is constantly evolving. Colonisation affected Indigenous peoples in different ways, and this influences the cognitive framings and relationship they have with the term. Indigenous peoples are found worldwide; we are from various backgrounds, nation states and cultures. For many, Indigeneity is a label that was put on us to signal an otherness in relation to the settler. My relationship with the term will always be problematic. The word identifies a collective of people who are experiencing colonisation, removal from Ancestral Lands, cultural genocide, foreign occupation and erasure from coloniser societies. But Indigeneity can also be a political state of mind, that requires an understanding of resistance, as well as an active movement towards freeing the land. Indigeneity is the experience of being colonised, which, because of capitalism, is still a daily occurrence. But most importantly here, Indigeneity is a set of knowledges and scholarships that provide successful blueprints for healing both Land and People. Our imaginations matter. We are still here. We deserve the dignity of time to imagine ourselves and the Land, away from supremacy structures, towards healing.

In freeing the Land, we give Nature the rightful platform to teach us. Trees can teach us about social cohesion and creating healing ecosystems. Beavers can teach us about nourishing complex ecosystems. For Nature to provide us with knowledges and imaginations, all living beings have to be left to flourish and exist.

All of us on this majestic planet have been given a sacred opportunity to experience life. This planet has given us everything we need to live in a healing future. We just have to let them be free.

We cannot heal until the Land is free.
We are not free until the Land is free.

12

WHAT THE WIND CAN BRING

Amanda Thomson

RECENTLY I'VE BEEN PAYING more attention to wildflowers. I've always been a birdwatcher, but flowers have added another dimension to my walks, slowing me, making me look down and around and up close as well as up and away. On steep hill walks, they're also an excuse to stop.

I'm an amateur and know little of their botany, aside from what's visibly before me. I only have a common, everyday knowledge of plants. As a child, I made daisy chains, plucked dandelions and blew their clocks to tell the time. I'd hold buttercups under my friends' chins to see if they liked butter. Some of my knowledge was more whimsical, other aspects were more practical. I somehow knew that if I was to get stung by nettles, I should reach for one of the dock leaves that always seemed to grow nearby to salve the sting. Yet I don't know if there's a reason why these plants grow in such close proximity (or whether they always do), or whether it was just a childhood perception. In recent years, I've come to be able to spot and name common plants like speedwells and sorrels, trefoils, clovers and willowherbs, and a few that are less common, like the wintergreens and, rarer still, twinflowers. I revel in names like self-heal, eye-bright, tormentil, butterwort, devil's bit scabious, bitter vetch.

We are learning more and more about the invisible and imperceptible concepts of entanglement and coexistence between plants and between species, above and below the soil. We've come to think of these subterranean connections as the *wood wide web*, but so many connections and inter-twinings are right in front of us. I know now that nettles also make for good soup and tea, and can work as a substitute for spinach. I've seen writhing masses of black caterpillars on a bed of nettles – those of the peacock butterfly. I know now that nettles are food for the larvae of red admirals and small tortoiseshell butterflies as well as moths like the burnished brass, and are an important plant for so many seed- and insect-eating birds. Other plants have other symbioses: in our garden blue butterflies flit around the bird's foot trefoil, and I know it's also the main food plant of the six-spot burnet moth, a beautiful, shimmery black-green moth with, yes, six scarlet spots on its wings. Common blues and meadow browns also love the common knapweed that grows alongside the trefoil, and siskins and goldfinches devour the seeds of that same plant at the end of the summer. Around and about throughout are countless flies, bees and micro-moths that I can't even begin to identify.

Our knowledge of the benefits and potential dangers of

plants is written into our folklore and histories: we know of their uses for medicine, poison and protection against evil, and of their economic importance too. St John's wort, used as an anti-depressant, was also said to ward off evil, butter-wort was used to curdle milk, and traditionally eye-bright was used for eye infections. Rowans were thought to protect people from harm, plantains stopped bleeding; self-heal used to heal wounds, stop bleeding, cure liver and kidney problems. So many plants were used to treat the coughs and colds and bronchial complaints that I can't help think would have accompanied Scotland's winter cold and damp. Heather was used for rope making and insulation, bent grass for thatching. Nettles used to be used as a cough medicine, a dye, for curdling milk, and, as fibres, woven into twine and fabric.

Our relationship to plants, however, has not always been and is not always benign. I remember hearing the art his-torian Anna Arabindan-Kesson talk about cotton, focusing on the plant as a way of exploring the interconnections of art, trade, colonial expansion and slavery. She discussed, among other things, 'negro cloth' – the rough cotton used for the clothing of the enslaved, as well as nineteenth-century depictions of Black bodies that were often central to aesthetic renderings of the cotton trade, and in doing so revealed the racialised discrimination and violence inher-ent to our relationship with this plant. Mark Nesbitt's and

Caroline Cornish's work on the 'seeds of empire: economic botany collections between nature and culture', shows how the material objects found in botany collections gathered during the nineteenth century reveal habits of appropriation and theft as well as scientific discovery. Botanic gardens are now often at the forefront of plant research and conservation, but we know their histories are tied to colonial expansion and exploitation, and they are filled with species brought back to the UK from the early 'exploratory' expeditions that were in the service of empire. Many of these plants are those that now populate the UK's gardens. The very terminology we use about plants seems loaded when we speak of native and non-native, 'invasive' and 'exotic' species, and in how some are seen as benign, some threatening. It's hard not to be conscious of language and this broader context when, in conservation and ecology, there's talk of protecting native species and removing invaders.

These days, I volunteer at a tree nursery, where we tend to the downy birches, montane willows and other species that, from impossibly small beginnings, will eventually find their way back into some of the upland wildernesses of the highlands where they will be replanted in areas where these species once grew. It is part of a project that seeks to restore such species to their former reach. One day, one of the other volunteers, an ecologist, mentions that the weeds I've been pulling out from around the birch seedlings are thale cress

– tiny white flowers that sit atop a five-inch-tall, slender stem with the tiniest of rosettes of leaves at its base. Thale cress, she tells me, was the first plant to be genetically sequenced, and it's odd to realise that this tiny and seemingly insignificant plant – which I've been plucking and discarding for hours – is actually quite important. It's all about context. As Richard Mabey memorably said, 'the ornamental in one place becomes the malign invader in another ... Just as readily the weed is domesticated into a food plant, or a children's plaything, or a cultural symbol.' When I get home I seek out more information and come across the scientist Nicholas Harberd's book *Seed to Seed*, which documents the life of thale cress and the journey to understand its genetic sequence. I'm not a scientist, and don't have that kind of brain, so in my reading only tangentially take in the intricacies and implications of the science of DNA that Harberd so eloquently describes, but he talks about the everyday in ways that send my thinking in new directions. For instance, I've never thought about the parts of a leaf before: 'The top half of the leaf is more concerned with the harvesting of sunlight during photosynthesis, whereas the bottom half is more for gaseous exchange, for the absorption of oxygen and carbon monoxide from the air.'

I start to look at leaves differently, lift them to see their undersides, and think of each leaf and what it needs to be capable of doing. I watch how things grow and senesce over the year. Even a single summer's day can hold so much change: I watch cat's ears and dandelions open up their flowers towards the sun and shift with its movement west, closing again in the evening to protect their seeds. And yet I know nothing of the nanosecond-to-nanosecond invisible adjustments and changes each plant, animal, living thing makes to adjust to the most micro of environmental changes. Harberd tells me there's a gene that's activated by a fall in temperature, that triggers another that encodes a protein that protects the thale cress's cells against the effects of cold weather. Its root uses gravity to know the direction to grow in, the cells that constitute it containing starch grains that are heavier and therefore pulled down by gravity towards the earth's core. And if a root encounters an obstacle – say, a stone – its tip triggers a hormone that sends a message back up the chain which will slow its growth, and allow it to feel its way around the obstacle. 'So a growing root is a sensitive thing, picks its way through the texture of the soil, finds the best path.'

Of course, plants, even without their practical, intrinsic value to us as food, clothing, shelter, medicine, help us think about speed and time through how they grow and transform over the changing seasons and years. I wonder how, if we paid even more attention to that aspect of them, that might root and connect us more. In early 2020, I visited the South African National Biodiversity Institute (SANBI) in Cape

Town as part of an Edinburgh International Book Festival Project called Outriders Africa. I went there to learn more about SANBI and the flora of Southern Africa, in particular that of the Cape Floristic Region and the Fynbos Biome. This is one of the most biologically diverse ecosystems in the world, and also a habitat under threat from climate change, increased urbanisation, the pressures of agricultural expansion and, of course, non-native species. SANBI is part of the Millennium Seed Bank Partnership, a project that started in 2000 to collect and safely store the seeds of the world's plants. 'With two in five plant species at risk of extinction', the MSBP notes, 'it's a race against time to protect our incredible plant life.' With a primary aim of conserving plants that have restricted ranges (and thus greater susceptibility to climate change), or are threatened in other ways, or are deemed to be 'useful', it now has 2.4 billion seeds stored in its vaults – including back-up 'insurance' stores of seeds from other partnership countries.

Part of SANBI's work is to collect the seeds of the country's indigenous wild plants, and I spent time with its scientists learning about what they do. I hunkered down among the Fynbos searching for seeds, and there was something very human about gathering them: how we moved, and sat in the fresh air, down among these shrubs and flowers, and how elements of experience, chance and luck coalesced into finding the seeds. Back in the lab, we'd sit on stools and carefully clean and sort them. I found there was a meditative quality to the work: it was something that couldn't be hurried. One of the scientists cut through one seed to check it, revealing a hollow shell, and it was the first time I'd looked at seeds under a microscope. He bisected another and I could see its layers, almost a juiciness, and with that, its viability, its possibility. I couldn't quite imagine the convoluted processes of development and growth that would allow its transformation from seed to plant, or the chain of events and circumstances that would enable it to get to the point where its cycle would start again. Or, beyond that, the multifarious benefits that it, or some of its kin, might provide to others. Something so everyday and so natural, but miraculous too.

One of the horticulturalists there told me of seeds from the Cape of Good Hope that were found in a 200-year-old notebook that had lain in the UK's National Archives (and the story of how they got there is again caught up in these complicated narratives that connect and divide). After they were found, scientists at Kew attempted to germinate them. They were unsuccessful with some, but one has become a healthy plant that grows in one of the glasshouses. If we are lucky, seeds can be hearty and resilient and will lie dormant until the circumstances are right. Nature can often find a way back in, if we give it time and the opportunity.

There are forests that have been planted in the last hundred years that started off essentially as monocultures

– stands of Norway spruce, Douglas fir, Scots pine, for example. Over time, with the offerings of birds that fly over and land within (acorns dropped, rowan and flower seeds excreted), and with what winds blow in from the adjoining lands, what is left from tyre tracks and what methods of planting bring in, they've become massively biodiverse ecosystems that can include the rarest of flowers and lichens and mosses. Nature can reclaim these places quickly, as it can abandoned buildings and old industrial sites.

In *Islands of Abandonment: Life in a Post-human Landscape* Cal Flyn writes about the shale bings of West Lothian in Scotland. These are unique remnants of a time in the latter half of the nineteenth and well into the twentieth century when Scotland was a massive extractor of oil, mining it from the shale beds found deep under the soil. In 2004, an ecologist, Barbra Harvie, surveyed them for their biodiversity and found over 350 plant species, including rare mosses and lichens, as well as locally rare animals and 47 species of birds with 30 breeding territories. There's Alpine clubmoss on one of the bings – a species more readily found in montane regions, and twayblades and common wintergreens grow, along with early purple orchids, greater knapweed, knotted pearlwort, crowberry. These and the other bings left after mining for coal or iron have become important islands, providing 'refuges for a wide range of animals and plants that are under increasing pressure in the surrounding area from farming and urban development'. What would the earth be like without humankind upon it, wonders Alan Weisman in *The World Without Us* – how might 'the rest of nature' be 'if it were suddenly relieved of the relentless pressures we heap on it and our fellow organisms?' I remember in particular the chapter about nature's reclamation of New York, how it would come from above as well as below – how seeds dropped from birds would land and catch and start to grow, how 'weeds' and trees would rip up through concrete and paving, how the water table would rise and flood the city.

Walking in Glasgow, I see how plants take their chances and establish themselves in the most precarious of footholds. Herb robert and enchanter's nightshade grow in the spaces between kerbstones, forget-me-nots blossom in the cracked surface of a pavement, mosses bloom in bright green lines up and down and along the seams of brick and stone walls. Trees push against and sometimes warp, sometimes engulf, railings planted when the trees were so much younger and slender. Over time their roots push up the pavements, creating small mounds and trip hazards. Rosebay willowherb, ox-eye daisies and umpteen kinds of grasses, dandelions and hawkweeds soften verges, rooftops and guttering. The birch tree is known as a pioneer species for its ability to flourish on disturbed, poor soils and environments, and I've seen it sprout from the most unlikely of places, including rooftops and in crevices between trunk and branch in older Scots pines.

Some plants survive within and sometimes even thrive on precarity; others are much more vulnerable to the ways we currently live. Some species have become indicators of our and the earth's well-being, and so much depends on how we value as much as what we value, and I think back to thale cress. When Jessica J. Lee was founding a journal of nature writing by people of colour, she named it the *Willowherb Review*, explaining, '*Chamaenerion angustifolium*, commonly known as rosebay willowherb or fireweed, is a plant that thrives on disturbed ground. Its seeds do well when transported to new and difficult terrain, so some – not us – may call it a weed.' I rather like that perspective and more: I see how we must appreciate plants, both common and rare. Some show resilience, perseverance, adaptation, survival; others a vulnerability and susceptibility to forces outwith their control that remind me of the precariousness of nature, as well as how it can bounce back. In our garden we have hawkbits, cat's-ears and dandelions (and I struggle to tell the difference between them), buttercups, trefoils, yarrow, clover, campion and vetches that grow as part of a wildflower meadow mix, and some might say our garden is just overgrown with weeds. There are other plants too that have arrived from somewhere else, or that have lain dormant in the ground to have sprouted again. Sometimes there's little difference between what's in the garden and what we see along the verges of nearby roads and paths and in the fields; it's just a bit more intense. Bees, moths, flies and butterflies flit on and in and around them, pollinating as they go. At the instant before some of the flowers turn to clocks and the breeze catches the seeds to disperse them, charms of goldfinches descend and pluck from the closed-up, dying and dead flowers, and fly off again. I wonder where they will go, and where the seeds they'll digest will disperse. I know that they'll be back for more, with siskins and maybe a redpoll or two, and again later on in the summer for the knapweed that will have flowered and gone, leaving its seeds as another food source for the birds.

NOTES

137   'Nettles used to be used . . .': William Milliken and Sam Bridgewater, *Flora Celtica: Plants and People in Scotland* (Birlinn, 2013).

     '. . . art historian Anna Arabindan-Kesson . . .': Her book, *Black Bodies, White Gold*, is published by Duke University Press (2021).

139   '. . . "the ornamental . . . a cultural symbol"': Richard Mabey, *Weeds* (Profile, 2010).

141   'SANBI is part of the Millennium Seed Bank Partnership . . .': www. kew.org/wakehurst/whats-at-wakehurst/millennium-seed-bank.

143   'Cal Flyn writes about . . .': in *Islands of Abandonment* (William Collins, 2021).

143 '... Barbıa Harvie, surveyed them ...': Barbra Harvie, *West Lothian Biodiversity Action Plan: Oil Shale Bings*, technical report, January 2005.

143–4 'What would the earth be like ...': Alan Weisman, *The World Without Us* (Picador, 2007).

145 '..."*Chamaenerion angustifolium*, commonly known as rosebay willowherb ...'": www.thewillowherbreview.com/about.

13

UPIRNGASAQ (ARCTIC SPRING)

Sheila Watt-Cloutier

I WRITE THIS FROM MY HOME in Kuujjuaq, an Inuit community in Nunavik, northern Quebec, Canada. We're located about 1,500 kilometres north of Montreal, on the tidal banks of the Kuujjuaq River, at a point where the northern extent of the treeline meets the Arctic tundra.

The remoteness of Nunavik has not entirely shielded us from the global reach of the current pandemic, and indeed outbreaks – although small in number – of infection have occurred in two of our communities. And so, for the past two months, I have been living in self-isolation, part of this time caring for my seven-year-old grandson, Inuapik. He's an extremely active little boy, always curious and observant. He has kept me on my toes from dawn to dusk.

It is now early June – the beginning of springtime in the Arctic, that brief period between winter and summer when life is miraculously renewed. The snow, apart from patches here and there, will soon vanish from the land. Our delicate plants, such as the purple saxifrage, fireweed and poppies, suddenly freed from their covering of snow, are quickly greening again. The snow buntings – *qupannuaq* – always the first to arrive, are being followed by flocks of other migratory birds, among them geese, ducks, loons and terns. The snow-white winter plumage of the ptarmigan – *aqiggiit*, our Arctic grouse – is taking on its summer camouflage. And our favourite fish, the Arctic char – *iqalukpik* – will soon begin their seaward migration from lakes connected to the upper reaches of the river, where they overwintered, to feed and replenish in the rich coastal waters of nearby Ungava Bay.

This is also a time when families look forward with intense joy to escaping community life for a while, heading to their traditional springtime camping spots near the mouth of the river or on the shores of Ungava Bay. Many of these sites have been occupied by the same Inuit families for generations, and being in any one of these places is to sense immediately the depth of history and connection they hold. In this way, year after year, families simultaneously renew their attachment to the land and to our ancestors. It is a time of storytelling, of remembering who we are. Here, our language, Inuktitut – ultimately a language of the land – reclaims its rightful place. And here our children, according to their age and gender, participate fully in traditional daily activities: learning and absorbing all the essential skills, aptitudes and attitudes required to survive and thrive on the land when their own time to be autonomous comes. In so many ways, the land never fails to invigorate and teach. Family and communal bonds are restored, and our spirits uplifted. We become healthier in mind and body, nourished by the 'country food' the land and sea provides. This includes a varied menu of goose and duck, fresh-run Arctic char and trout and, of course, *natsiq*, the common seal, a staple food of Inuit coastal dwellers everywhere. This ample diet is inevitably supplemented by seagull, goose and eider duck eggs, gathered from islets just off the shore. At low tide we dig for shellfish, mostly mussels, or catch sculpins, a small, spiny fish we call *kanajuq*, stranded in rocky pools by the falling tide. Raw, crunchy seaweed, gathered from these same pools, occasionally complements the boiled *kanajuq*.

With the signs of spring all around me, and my dreams of soon being able to get out on the land again, in season to go berry picking with fellow Inuit women, it's perhaps not surprising that my thoughts have turned to the place of nature in Inuit life. In our language we have no word for 'nature', despite our deep affinity with the land, which teaches us how to live in harmony with the natural world. The division the Western world likes to make between 'man and nature' is in the traditional Inuit view both foreign and dangerous. In Western thinking, humans are set apart from nature; nature is something to strive against, to conquer, to tame, to exploit or, more benignly, to use for 'recreation'.

By contrast, Inuit place themselves within, not apart from, nature. This 'in-ness' is perfectly symbolised in our traditional dwellings of the past: *illuvigait* (snow houses) in winter and *tupiit* (sealskin tents) in summer. What could be more within nature than living comfortably in dwellings made of snow and sealskin!

This is especially true of our relationships with the animals that sustain us: the *puijiit* – sea mammals – seals, whales and walruses; and the *pisuktiit*, the land animals, in particular caribou and polar bear. No other people have relied so exclusively on animals as my Inuit ancestors.

In one of the world's harshest environments, these Arctic animals provided everything needed to sustain human life. Their flesh supplied all the nutrition required for a healthy diet. From their skins, cut and worked as needed, clothing and shelter were sewn. The blubber of marine mammals fuelled the *qulliit* – our soapstone lamps – providing light and a little warmth for the snow houses in the depths of winter. From bones, ivory and caribou antler, tools, utensils and hunting equipment were expertly fashioned. Thread, strong and waterproof, used with the seamstresses' delicate bone and ivory needles, came from the sinews of caribou and beluga whales. The reliance on animals was total. Other than berries and roots, in some places available at the end of the Arctic's brief summer, there was no plant life, no agriculture, to fall back on should the hunt fail.

Our ancient beliefs held that the animals we relied upon had souls, just like ours, which needed to be treated with respect and dignity. In the early 1920s, Avva, an Inuit shaman from Igloolik, whose descendants I know well from my residential schooldays, as well as from the time I lived in Iqaluit, Nunavut, for almost twenty years, famously summed up these beliefs at the very core of our pre-Christian identity: 'All the creatures that we have to kill and eat, all those that we have to strike down and destroy to make clothes for ourselves, have souls, like we have, souls that do not perish with the body, which must therefore be propitiated lest they should revenge themselves on us for taking away their bodies.'

Founded on respect, our appeasement of the animals we harvested took many forms: for instance, giving a newly killed seal or walrus a mouthful of water, a practice based on the knowledge from a deep understanding of and connection to the animals we hunt that these mammals, having spent all their lives in the sea, craved a drink of fresh water. Taboos associated with particular animals were strictly observed. In this way, care was taken to avoid mingling creatures of the sea with those of the land, and so there were prohibitions against sewing caribou-skin clothing on the sea ice. Nor could the flesh of seal and caribou be boiled in the same pot. I remember my mother reminding me of this even when I would eat both frozen fish and frozen caribou together. Above all, the absolute bond between my ancestors and the animals they hunted (and, by extension, the land, sea and air) was founded on respect. Hunters never boasted about their prowess. Abusing animals in any way, or mocking them, or using them for 'sport', resulted in serious consequences for society, as did disputes over sharing. In response to maltreatment or insults, animals would withdraw from hunting grounds. Hunters were obliged to kill only animals who 'presented' themselves for the taking. This is exactly why, when I lived in the south and made visits home to Kuujjuaq in the early spring, and we hunted *aqiggiit*, my mother would say to me: 'Isn't it wonderful that the *aqiggiit* brought themselves to you so that you could take them back with you to eat in Montreal!' My mother always had that deep Inuit understanding of how life gives life.

There's an ancient tale that vividly illustrates the ethical imperative for Inuit of respecting animals when they 'present' themselves, a story that explains why walruses disappeared from a place called Allurilik, a large inlet on Ungava Bay, just over 200 kilometres north-east of my home in Kuujjuaq. It is said that here there was once a hunter out on his *qajaq* (kayak) looking for walruses. Suddenly, a small walrus surfaced in front of him and begged to be taken because it craved a drink of fresh water. Noticing that this little walrus had very small, deformed tusks, the hunter refused, saying: 'Go away . . . I don't want you. Your tusks are too small and deformed!' Hearing these words, the walrus was deeply offended and went away. Shortly after that incident, all the other walruses left the area and never came back. It is said that the caribou, after hearing about the insult, also abandoned the land around Allurilik. The lesson here is that all animals presenting, or in my mother's words 'bringing' themselves to the hunter, should be understood not as confirmation of death, but affirmation of life.

Indigenous communities and cultures everywhere have been ravaged by contact with the Western world. Introduced diseases, against which they had no resistance, decimated their populations. Christianity – usually the forerunner of colonialism – pushed aside Indigenous belief systems, altering the way they viewed the world, and endangering their mutual bonds with nature, with the land, animals and forests that sustained them.

Europeans first came into contact with my Inuit ancestors on the south shore of Ungava Bay just over 200 years ago. From that moment forward, our essential oneness with the natural world was challenged and would eventually change forever. Like the start of any infection, at first the symptoms were subtle. In those early days of contact, the Arctic, in the European imagination, offered nothing worth exploiting. Our land was dismissed as a barren wilderness, covered in snow and ice for most of the year, inexplicably inhabited by a few nomadic 'heathens'. Above all, the Arctic, with its ice-filled summer seas, was seen as a sort of adversary to be heroically conquered in Europe's futile efforts to find a northwest passage to the 'riches of the Orient'.

Regardless, wherever Europeans 'discovered' Indigenous peoples, commerce and Christianity were sure to follow, and my Arctic homeland was no exception. In time, the inescapable reach of the Europeans extended to our shores. We named them 'Qallunaat'. Men of the Hudson's Bay Company were the first to arrive, setting up, in 1830, a trading post on the east side of the Kuujjuaq River, more or less across from the place where the modern community of Kuujjuaq now stands. Shortly after the turn of the century, an Anglican mission was also established there, joined by a Catholic mission in 1948.

We slowly began to accept these strangers in our land, and over time we gained some understanding of their ways. But through coercion, when our own powerful spiritual beliefs, which included shamanism, drum dancing and throat singing, were forbidden and considered 'taboo', our people eventually converted. The traders' goods were an obvious convenience, especially metal items such as needles, knives, kettles, traps and firearms, joined later by an increasing selection of woven fabrics, sewing materials and basic foodstuffs, including flour, lard, sugar and tea. And, of course, tobacco. Although we could not have known it at the time, the seeds of consumerism, profound and dangerous changes to our diet and new diseases were unobtrusively planted among us. We distanced ourselves from the Qallunaat, and our interactions with them tended to be irregular and infrequent. We continued to live on the land, moving predictably from place to place in harmony with the animals, which had sustained us for countless generations. From time to time, usually travelling by dog team, visits were made to the post to trade furs, or to celebrate Christmas at the mission. Yet despite this distancing, our way of life, our unity with nature, was to change forever. Our traditional perception of time, for example, which had ticked to nature's clock – the rising and setting of the sun, the position of the stars, the cycle of the tides, the succession of the moon months – now needed to make room for the Christian calendar. Suddenly there was a unit of time called a 'week'; how very strange the idea must have seemed to my ancestors that one in every seven days was a special day when hunting and all other 'work' had to stop! Similarly, the traders' constant need for fur, especially white fox, began to alter our subsistence patterns as we spent increasingly more time on our traplines during winter.

Throughout this initial period, which lasted from the mid-1930s to the late 1950s, of coming to terms with the now permanent presence of traders and missionaries in our lands, our lives remained relatively unchanged. We continued to live in extended family groups, distributed along the coast of Ungava Bay. Our culture, values and traditions remained strong, as did our language, which easily incorporated new concepts and objects brought from the south. 'Sunday', for example, we called *allituqaq*, literally a time when we have to 'respect a taboo' – in this case the taboo against hunting on that particular day. And the kettles and pocket knives we bought from the traders were named *tiqtititsigutik* or *uuju-liurutik* (that which is used to boil something) and *puutta-juuq* (that which regularly unfolds). So we slowly adapted to the newcomers, integrating their ideas and material things at our own pace. In the beginning we came to view this new relationship with the Qallunaat world as essentially balanced and sustainable. Above all, by continuing our life on the land, usually several days of dog-team travel away from the Qallunaat dwellings, we were able to retain our autonomy over the aspects of our lives that mattered most. This included our bonds with the land (including the sea ice) and its animals; and, most important of all, teaching our children the traditions, philosophies and skills needed to continue this land-based life.

Looking back on this period we certainly did not think that this way of life would last forever. And indeed, it didn't. In the 1950s and early 1960s the Canadian government suddenly took an interest in 'its territories' in the far north. Focus

on the area first came from the construction of the so-called
Distant Early Warning Line, a sort of necklace of defence
radar stations built by the US military above the Arctic
Circle, from Baffin Island, Canada, to Wainwright, Alaska.
With advancing technology and increasing explorations by
prospectors, mineral exploitation in the Arctic was becoming a real possibility. And there were also tragic reports of
inland-dwelling Inuit in Canada's 'barren grounds' starving
to death. The Canadian government decided it was time to
act. Without any meaningful consultation, they instigated a
policy to move Inuit from the land into settlements that, in
most instances, would be built at sites previously established
by the Hudson's Bay Company and the missionaries.

From the start, the government's policy to move us 'off
the land' was misguided and paternalistic. The idea was to
make the 'administration' of Canada's Eskimos (as we were
then called) easier. We were seen as a problem needing to
be fixed. This would be mended by gathering us into settlements, building houses for us and 'educating' our children
in English with a 'Dick and Jane' curriculum, an education
that had nothing to do with what we knew to be the real
world. We would partake of the government's assistance
programmes such as family allowances (which sometimes
could be withheld if we didn't send our children to school)
and, when needed, social assistance payments and subsi-
dised housing. Along with the provision of health services,

these seemingly positive enticements were difficult to resist. Nowadays we recognise these offerings as coercive, though strangely packaged in well-meaning wrappings.

In my case, our family's move into the settlement happened in 1957, earlier than for most Inuit then living in the Canadian Arctic. At the time, we were living at Old Fort Chimo, where I was born, and where the Hudson's Bay Company still ran a trading post. Across the river from us, the US military had built a weather station and landing strip during the Second World War, one of several airstrips on a northern route to Europe, along which the Americans used to ferry aircraft to Britain. After the war the US transferred the site's buildings and airstrip to the Canadian government and in time, under its 'ingathering of Inuit' policy, these became the present-day community of Kuujjuaq.

With the move, things happened very quickly. At first, we expected that this new world in which we suddenly found ourselves would be as wise as our own. But it wasn't. It turned out that our new world was deeply dependent on external political and economic concepts and forces utterly at odds with our ways of being. In particular, its structures seemed to have nothing to do with the natural world. Almost immediately, we started to give away our power. For a while we thought that if we were patient – as the Inuit hunters necessarily are – that patience would pay off. But we soon lost that sense of control over our lives, especially

April 1956, in Old Fort Chimo, where the author was born, across the river from where Kuujjuaq is now. Back row, left to right: Maggie Gordon, Kitty Munick, Mary Simon and Bridget Watt, the author's sister. Front row: Joanne Ploughman, Donna Ploughman and the author.

over the upbringing of our children. They were brought into the classrooms of southern institutional schooling, a concept totally foreign to us, where they were given an 'education' that had nothing to do with the knowledge and skills we needed for life on the land. All our traditional character-building teachings went out the window, and our social values began to erode. When we surrender our personal autonomy, we also give away our sense of self-worth, we lose the ability to define ourselves and to navigate our own lives. Being brought into the settlements was the beginning of the end for our traditional way of life. In the settlements we lived in a kind of bubble, separated from the natural world, exchanging our independence for increasing dependency.

In this new, confusing life – which, at least on the surface, seemed to meet all our basic needs – we also lost, above all, our sense of purpose. In our attempts to replace this loss with something else, many of us drifted into addictions and self-destructive behaviours, made worse by unemployment and poverty. This downward trend has played out over several generations in the most horrific ways, seen most tragically in the current levels of suicide among Inuit youth.

I was in my late teens when we experienced our first suicide in Kuujjuaq, a young Inuit woman, though she was not actually from our community. Traditionally suicide, in Inuit society, was rare and affected mostly adults, so this was shocking and incomprehensible to us all. Nowadays it's a tragic fact that our Inuit youth suicide rates are among the highest in the world. I have no doubt whatsoever that this tragedy is rooted in our move from the land, and the subsequent erosion of our culture and values, not to mention the historical traumas of forced relocations, the slaughter of our sled dogs and abuse in many forms by those with authority. Whatever the underlying causes, these suicides can often be impulsive. In our traditional ways, impulsivity had no place. On the land, to act impulsively was to put yourself and everyone else around you at risk. Even under extreme pressure, decisions had to be weighed carefully. In our upbringing we were taught to develop that sense of holding back, of reflecting and being focused: our very lives depended on us avoiding any urges towards reckless behaviour.

Along with many others of my generation, I was fortunate enough to have spent my formative years deeply steeped in Inuit traditional ways and values that gave us our understanding of the world and our place in it and, importantly, our responsibilities to it. My age group still talk about this – that sense of training and the grounding we got, which have kept us going and made us resilient.

My early years in Kuujjuaq cocooned me in these traditions, thanks, primarily, to two incredibly strong women: my mother and my grandmother. I also learned by observing my uncle, a skilled hunter and community leader with a lot of integrity and dignity, as well as my older brothers, who had been taught many skills by my uncle and other men in our community. Beyond these, teachers enough in themselves, were the always gently instructive social interactions I enjoyed with the small community around us. This supportive and caring circle was occasionally enriched by Inuit visitors from other parts of Ungava Bay, coming into Kuujjuaq to trade, travelling by dog team in the winter or canoe in the summer. To this day I can vividly recall their words as I sat, silent and wide-eyed with amazement, listening to them relate their news and stories to my mother and grandmother.

Of course, these occasions were always an opportunity to liberally share in whatever country food we had at hand. Depending on the season, this could be any combination of fish, ptarmigan, seal and the choice parts of caribou, raw, dry, frozen or cooked, according to preference. Most often these foods would be enhanced by our traditional condiment, a dipping sauce we called *misiraq*, made from fermented seal oil. Sharing the food our land provides is a deeply held Inuit tradition, indeed an imperative – there's no other word for it. Wherever we are, this practice is still at the core of our family and community life. In this unspoken ritual, sharing nature's bounty renews, again and again, our bonds with each other and the land that sustains us.

I have an early memory that brought all these strands together, underscoring our essential place within nature that I didn't fully understand at the time. Inuit have many categories of relationships and relationship terms without an exact equivalent in the Western world. Traditionally, personal names given at birth were said to carry souls, and they immediately established a wide network of relationships, even mutual responsibilities, often extending beyond the immediate family. Nor were personal names ever gendered. For instance, a baby boy named after, say, his maternal grandmother would be addressed by his own mother as *anaana* – meaning mother – and, in some cases, at least until puberty, would be dressed and even socialised as a girl. Family members would notice with delight how he took on some of his grandmother's personality traits and mannerisms. In this way, his grandmother continued to live through him.

A particularly significant relationship, in terms of linking community and nature, was initiated at birth with the person who cut the umbilical cord, usually a woman. If the baby was a girl, this woman would be known as her *sanajik*; if a boy, she would be his *arnaqutik*. The baby then became the *arnaliak* of her *sanajik*, or the *angusiak* of his *arnaqutik*.

Both my grandmother and mother were known for their midwifery skills, and so they had a good number of *angusiaks* and *arnaliaks*. One of the main obligations of their *angusiaks* was to present them with their first catch from the hunt – be it fish, seal, ptarmigan or caribou, a rite of passage, celebrating the very foundations of Inuit society: that is the sacred, interdependent relationship between the animals we hunt and our hunters. When I was a small girl, I saw this ritual played out many times as these budding hunters – my grandmother's *angusiaks* – honoured their obligations to her. One at a time, every other month or so, young men would come by our house to present their catch. In response my grandmother put on an amazing performance. This normally quiet, dignified elderly woman would suddenly

Left to right: Daisy Watt (the author's mother), the author and their friend Pasha Simigaq in Kangirsuk, Nunavik, c. summer 1957.

turn into an animal-like person, rolling around and making animal noises on the floor. Sometimes she would nibble the young hunter's hand or wrist, acknowledging their power, encouraging him to become a great hunter. I watched this startling performance almost in embarrassment because then, as a child, I didn't fully grasp its deep ceremonial significance, beyond sensing it was a necessary part of our hunting culture.

For their part, the girls and young women who were my grandmother's and mother's *arnaliaks* would be similarly honoured and encouraged when they brought gifts demonstrating their increasing ability in sewing. Proper, well-made skin clothing, warm and watertight as needed, was an absolute necessity for the successful provider. Inevitably my grandmother's ritual would finish with the young men or women we had just celebrated leaving the house confident and reassured, knowing that their work or hunt had been well received, endorsed by the woman who had helped to bring them into this life.

Sadly, this ancient custom is not much practised now, though, from my early years working at the Kuujjuaq nursing station, I assisted at several births and therefore have my own set of *angusiaks* and *arnaliaks*. I do my best to keep up with them, encouraging them over the years as they successfully fulfil their varied roles in life. Some of the young men have brought me their first hunt; the young women, gifts of their picked berries, caught fish or first pieces of handiwork. My response was not as dramatic as my grandmother's, but it was no less full of delighted gratitude. I was humbled and honoured that they had thought to keep this tradition alive.

Despite the extensive damage done to Inuit society and culture when we moved from the land into the villages, there is, in most of these settlements, an essential core of families instinctively committed to maintaining our traditions. Individual members of these families, even while living within the semi-urban settings, strive to relate to the land and its resources in the same respectful way that sustained us prior to the move. They acquire an intimate knowledge of their local area and the various animal species it supports. The men employ many of the same hunting skills used in former times, while the women prepare and soften the skins of seals and caribou for the clothing they make for themselves and their hunters, using techniques, patterns and stitches handed down by an endless succession of mothers, aunts and grandmothers. Most importantly, members of these families embody the essential philosophies and understandings of the land and animals that enabled us to thrive over countless generations before we suffered the consequences of European contact. In a real and substantial sense such Inuit keep the vital flame of our culture alive. They are an irreplaceable resource, in both practical and intellectual ways, and they need and deserve every possible means of support.

But beyond the challenges this already vulnerable way of being endures, in the face of the Arctic's rapidly increasing urbanisation (and globalisation), there is another imminent threat – no less insidious – that, unless checked, will end forever our unique attachment to the land and its life-giving resources: climate change.

Dramatic climate change caused by greenhouse gases has left no feature of our Arctic landscape, seascape or way of life untouched. Climate change now threatens our very culture, our ability to live off the land and eat our country foods. Nowhere else in the world are ice and snow so essential to transportation and mobility. And yet the snow and ice coverings over which we access our traditional foods are becoming more and more unreliable and therefore unsafe, leaving our hunters more prone than ever to breaking through unexpectedly thin ice or being swept out to sea when the floe-ice platform on which they are hunting breaks off from the land-fast ice.

Additionally, climate change is affecting the migration patterns and routes of the animals we rely upon. This means that our hunters have to travel further, often over unsafe and unfamiliar trails, to access our country food. So, when we can no longer count on our vital, long-established travel routes, and can no longer find the animals where they should be, the matter immediately becomes an issue of safety and security at several levels.

With less traditional food available, many families are forced to shift away from our traditional food to a far less healthy diet shipped from the south, consisting mainly of processed foods crammed with sugar, salt and carbohydrates. It is no surprise that in Canada our Arctic communities are experiencing rapidly rising rates of diabetes and other food-related illnesses, trends that will only continue as we move away from a country-food diet.

Hand in hand with climate change is the ongoing threat of Arctic resource development targeting our rich mineral and oil deposits. Our anxieties on this front are regularly dismissed by our own governments, who see the Arctic as the next super-energy 'feeder' for the world. In the greater scheme of things, Inuit concerns over their livelihoods and environment are dismissed as unimportant.

As someone who has led pioneering global work on connecting climate change to human rights, I am convinced that the escalating pressures we now face regarding resource development will deepen the need for all parties to adopt a rights-based approach in the search for solutions to these problems.

Everyone benefits from a frozen Arctic. The future of the Arctic environment, and the Inuit it supports, is inextricably tied to the future of the planet. Our Arctic home is a barometer of the planet's health: if we cannot save the Arctic, can we really hope to save the forests, the rivers and the farmlands of other regions?

We can also no longer separate the importance and the value of the Arctic from the sustainable growth of economies around the world. In the international arenas, where I have personally been involved, the language of economics and technology is always calling for further delays on climate action. We are constantly reminded that making any significant efforts to tackle greenhouse gas emissions will negatively impact the economy. But I truly believe that we must reframe the terms of the debate regarding the implications of environmental degradation, resource development and climate change in the Arctic and move beyond relying solely on the language of economics and technology. What is needed is a debate emphasising human and cultural rights. Focusing only on economics and technology separates the issues from one another as opposed to recognising the close connections among rights, environmental change, health, economic development and society. Ultimately, addressing climate change in the language of human rights and building the protection of human rights into our global climate agreements are not just matters of strategy, but moral and ethical imperatives that require the world to take a principled and courageous path to solve this great challenge.

And I strongly believe that we need to reimagine and realign economic values with those of the Indigenous world, the Inuit world, rather than merely replicating what hasn't worked with the values of Western society. And who better than the Inuit themselves, who are natural conservationists, to be out there on the land and ice as paid guardians and sentinels? How deeply affirming that would be for our hunters, whose remarkable traditional knowledge is so undervalued. What better way to reclaim what was taken from us: our pride, our dignity, our resourcefulness, our wisdom? We don't want to just be victims of globalisation. We can offer much more to this debate if we could be included on every level. We have lived through states of emergency for decades now and we have attempted to signal to the world the climate crisis looming in front of us. Sadly, many in other parts of the world who are now experiencing these states of emergency, with the loss of their homes and livelihoods to fires, floods and other unnatural disasters caused by climate change, are finally beginning to see the connections.

So my message to you is: look to, listen to and support morally, respectfully, openly and, yes, financially the Inuit world, the Indigenous world, which from a place of deep love for its culture and traditions is fighting for the protection of a sustainable way of life. Not just for its peoples themselves, but for all of us. Heed and support those voices and their aspirations. We will help guide you as we navigate through these precarious situations together. Don't be on a mission to save us: this is not what we want or need. But together in equal partnership, with an understanding of our common humanity, we can do this together.

EPILOGUE

I began writing this piece in my Arctic home in Kuujjuaq. I am still here; still following the recommended social distancing and self-isolation measures brought on by the pandemic, a grim reminder of how interconnected and interdependent we all are. The remoteness of the Arctic no longer sets us apart from the rest of the world. This pandemic has also helped to break open unresolved issues of social injustices and racism: North American and European countries that often tout their great human rights reputations are now being fully exposed for outdated racist policies and attitudes which undermine, and put at great risk, the health of those most vulnerable. Black communities, the American Indian nations and our own Indigenous populations here in the Arctic show clearly that the economic and health gaps are huge in comparison to the white populations of rich countries. The only difference that sets us in the Arctic apart from our Black and Navaho brothers and sisters with the human losses they have suffered is that to date, as I write this, the geographic distance at which we live and the lockdown of flights coming in and out of our regions have thus far protected us. That could easily change overnight if and when there is a resurgence of the virus. History has shown us that many Inuit families were wiped out by past epidemics, so our leaders are extremely committed in their attempts to keep the virus out of our regions.

In many different ways, the pandemic has also given us pause. I have been taking the time to use this pause as a gift to reflect on new possibilities, new perspectives. The world should not, cannot, go back to business as usual without a clearer understanding and consciousness of how we live.

In my life's work, dealing with climate change and the protection of our Inuit way of life, I have often wondered what is going to eventually 'give'? What big event will finally wake us up to the realisation that the reckless, damaging way in which we do business around the world is unsustainable? In my talks, I often ask: 'What will it take to get the health back into our atmosphere so the earth can start to heal?' The earth is a living, breathing entity. If we care for it, it will heal just as our bodies do when we are sick.

I have always sensed the earth would reach its limits soon enough, but I didn't realise it would be in the form of a deadly virus that would virtually halt (at least temporarily) so many of our unsustainable activities. Almost immediately, the air and the waters of the world's industrial cities began to clear. Animals, suddenly relieved from unwelcoming human activity, appeared in some deserted city streets, as if reclaiming their rightful space. Nature is resilient, if only given the chance. Let's pay heed to these lessons. Let's make this a time of seeing that human trauma and planet trauma are one and the same. Let's not wait for another virus, driven by climate change and environmental degradation, to terrify us, too late, into half-hearted action. There is no time for half measures. The values and knowledge of the Indigenous world, the survival of which utterly depends upon living within nature, not apart from nature, hold the answer to many of the global challenges we face today. Indigenous wisdom is the medicine we seek in healing our planet and creating a sustainable world. I truly believe this.

14

BRAIDING SWEETGRASS

Robin Wall Kimmerer

HOLD OUT YOUR HANDS and let me lay upon them a sheaf of freshly picked sweetgrass, loose and flowing, like newly washed hair. Golden green and glossy above, the stems are banded with purple and white where they meet the ground. Hold the bundle up to your nose. Find the fragrance of honeyed vanilla over the scent of river water and black earth and you understand its scientific name: *Hierochloe odorata*, meaning the fragrant, holy grass. In our language it is called *wiingaashk*, the sweet-smelling hair of Mother Earth. Breathe it in and you start to remember things you didn't know you'd forgotten.

A sheaf of sweetgrass, bound at the end and divided into thirds, is ready to braid. In braiding sweetgrass – so that it is smooth, glossy, and worthy of the gift – a certain amount of tension is needed. As any little girl with tight braids will tell you, you have to pull a bit. Of course you can do it yourself – by tying one end to a chair, or by holding it in your teeth and braiding backwards away from yourself – but the sweetest way is to have someone else hold the end so that you pull gently against each other, all the while leaning in, head to head, chatting and laughing, watching each other's hands, one holding steady while the other shifts the slim bundles over one another, each in its turn. Linked by sweetgrass, there is reciprocity between you, linked by sweetgrass, the holder as vital as the braider. The braid becomes finer and thinner as you near the end, until you're braiding individual blades of grass, and then you tie it off.

Will you hold the end of the bundle while I braid? Hands joined by grass, can we bend our heads together and make a braid to honor the earth? And then I'll hold it for you, while you braid, too.

I could hand you a braid of sweetgrass, as thick and shining as the plait that hung down my grandmother's back. But it is not mine to give, nor yours to take. *Wiingaashk* belongs to herself. So I offer, in its place, a braid of stories meant to heal our relationship with the world.

ACKNOWLEDGEMENTS

'Before Roots' is a revised extract from *Entangled Life: How Fungi Make Our Worlds, Change Our Minds, and Shape Our Futures* by Merlin Sheldrake, published by Vintage (merlinsheldrake.com).

'Plants Know' first appeared in *Purple Magazine* (purplemagazine.fr).

'*Upirngasaq* (Arctic Spring)' was first published in *Granta* 153: Second Nature (Autumn 2020), guest-edited by Isabella Tree and published by Granta Trust.

The extract from *Braiding Sweetgrass* was first published in the UK by Penguin Random House.

Chapter opening illustrations: *Photosynthetics* (*Fotosintéticos*) © Eduardo Navarro, 2021. Commissioned by Wellcome Collection and La Casa Encendida, in partnership with Delfina Foundation. Photographs courtesy of the artist Gustavo Lowry.

'Wilder Flowers' illustrations © Rowan Hisayo Buchanan.

'*Upirngasaq* (Arctic Spring)' photographs courtesy of
   Sheila Watt-Cloutier.

Amanda Thomson acknowledges SANBI for hosting her
in Cape Town and to the Edinburgh International Book
Festival for the opportunity to be part of the Outriders
Africa project.

Epigraph to 'Strange Soil', p. 29 © Wendell Berry, *The
   Unsettling of America: Culture and Agriculture* (1977;
   revised edition, 1996).
With thanks to Teresa Orbegoso for the use of her words
   (p. 77) from 'Perú', *Poemas de Teresa Orbegoso* (Buenos
   Aires Poetry, 2016).

ABOUT THE AUTHORS

Eduardo Navarro was born in Argentina in 1979. His recent solo shows include: (breathspace) at Gasworks, London, UK (2020); Into Ourselves at Drawing Center, New York, US (2018); and OCTOPIA at Museo Rufino Tamayo, Mexico City, Mexico (2016). Some of his most recent works were included in group shows such as: Museo Nacional Thyssen-Bornemisza, Madrid, Spain (2020); SITE Santa Fé Biennial (2018); and the São Paulo Biennial (2010 and 2016).

Michael Marder is Ikerbasque Research Professor in the Department of Philosophy at the University of the Basque Country (UPV-EHU), Vitoria-Gasteiz, Spain. His writings span the fields of ecological theory, phenomenology and political thought. He is the author of numerous scientific articles and eighteen monographs, including *Plant-Thinking* (2013); *Phenomena—Critique— Logos* (2014); *The Philosopher's Plant* (2014); *Dust* (2016); *Energy Dreams* (2017); *Heidegger* (2018); *Political*

Categories (2019); *Pyropolitics* (2015, 2020); *Dump Philosophy* (2020); *Hegel's Energy* (2021); and *Green Mass* (2021) among others. For more information, consult his website michaelmarder.org.

Merlin Sheldrake is a biologist and a writer. He received a PhD in Tropical Ecology from Cambridge University for his work on underground fungal networks in tropical forests in Panama, where he was a predoctoral research fellow of the Smithsonian Tropical Research Institute. He is a musician and keen fermenter. *Entangled Life*, his first book, was an international bestseller, winner of the Wainwright Prize 2021, and has been nominated for a host of other prizes, including the British Book Awards Book of the Year 2021 for Narrative Non-Fiction and the Rathbones Folio Prize 2021.

Abi Palmer is an artist and writer exploring the relationship between linguistic and physical communication. *Crip Casino* – her interactive gambling arcade parodying the wellness industry and institutionalised spaces – has been exhibited at Tate Modern, Somerset House and the Wellcome Collection. Her debut book *Sanatorium* (Penned in the Margins, 2020) is a fragmented memoir, jumping between luxury thermal pool and blue inflatable bathtub.

Rebecca Tamás is the author of the poetry collection *WITCH* (Penned in the Margins, 2019) and the essay collection *Strangers: Essays on the Human and Nonhuman* (Makina Books, 2020). She is a Lecturer in Creative Writing at York St John University, where she co-convenes the York Centre for Writing Poetry Series.

Emanuele Coccia is an Italian philosopher, academic and writer. He is Associate Professor at the École des Hautes Études en Sciences Sociales in Paris, and the author of *The Life of Plants: A Metaphysics of Mixture.*

Rowan Hisayo Buchanan is an American and British novelist, short story writer and essayist. She received her BA from Columbia University and her MFA from the University of Wisconsin-Madison. Her first novel *Harmless Like You* was a *New York Times* Editors' Choice and an NPR Great Read. In the UK it won the Authors' Club First Novel Award, a Betty Trask Award, and was shortlisted for the Desmond Elliot Prize. Her work has appeared in *Granta*, *Guernica*, the *Guardian*, and the *Paris Review*, among other places.

Kim Walker is an author, researcher and PhD candidate undertaking a project on the *Cinchona* collections at the Royal Botanic Gardens, Kew. She is funded through a

TECHNE AHRC NPIF Doctoral Training Partnership at Royal Holloway, University of London. Her research interests are in medicinal plants and ethnobotany.

Nataly Allasi Canales, originally from the Peruvian Amazon, is a postdoctoral researcher at the University of Copenhagen. After her MSc in Bioinformatics in China, she received a PhD in Evolutionary Genomics in Denmark for her thesis on the fever tree using multidisciplinary approaches including genomics, biochemistry, traditional knowledge and archives. She is interested in continuing working on neglected Amazonian and Andean organisms using genomics and biochemistry to validate traditional knowledge.'

Jessica J. Lee is a British-Canadian-Taiwanese author and environmental historian, and winner of the Hilary Weston Writers' Trust Prize for Nonfiction, the Boardman Tasker Award for Mountain Literature and the RBC Taylor Prize Emerging Writer Award. She is the author of two books of nature writing: *Turning* (2017) and *Two Trees Make a Forest* (2019), which was shortlisted for Canada Reads 2021. She has a PhD in Environmental History and Aesthetics and was Writer-in-Residence at the Leibniz Institute for Freshwater Ecology in Berlin from 2017 to 2018. Jessica is the founding editor of the *Willowherb Review* and a researcher at the University of Cambridge. She lives in London.

ABOUT THE AUTHORS

Sumana Roy is the author of a work of non-fiction, *How I Became a Tree*; *Missing: A Novel*; *Out of Syllabus: Poems*; and *My Mother's Lover and Other Stories*, a collection of short stories. She is Associate Professor of English and Creative Writing at Ashoka University.

Susie Orbach is a psychotherapist, psychoanalyst, writer and social critic. She is the founder of the Women's Therapy Centre of London, a former *Guardian* columnist and visiting professor at the London School of Economics and the author of a number of books including *What Do Women Want*, *On Eating*, *Hunger Strike*, *The Impossibility of Sex*, *In Therapy*, *Bodies* – which won the Women in Psychology Prize – and the international bestseller *Fat is a Feminist Issue*, which has sold well over a million copies. *The New York Times* said, 'She is probably the most famous psychotherapist to have set up couch in Britain since Sigmund Freud.' She lives in London and lectures extensively worldwide.

Araceli Camargo is a cognitive neuroscientist. She is the lab director and lead scientist at the Centric Lab (in partnership with UCL) focused on the relationship between health and place. Araceli is a descendant from Turtle Island Peoples and holds an MSc in Cognitive Neuroscience from King's College London.

Amanda Thomson is a visual artist and writer and a lecturer at the Glasgow School of Art. Her creative non-fiction has appeared on BBC Radio 4, and in a number of anthologies. She also writes for the *Guardian*'s 'Country Diary'. She earned her doctorate in interdisciplinary arts practice, focusing on the landscapes and forests of the north of Scotland, in 2013, and her work is frequently about the Scottish Highlands. She lives and works in Strathspey, in the Highlands, and in Glasgow. Her first book, *A Scots Dictionary of Nature*, is published by Saraband Books.

Sheila Watt-Cloutier is an Inuk environmental, cultural and human rights advocate, and the author of *The Right to Be Cold*. She has been nominated for the Nobel Peace Prize and has received the UN Champion of the Earth Award, the Sophie Prize, the Jack P. Blaney Award for Dialogue and the Right Livelihood Award.

Robin Wall Kimmerer is a mother, scientist, decorated professor and enrolled member of the Citizen Potawatomi Nation. She lives in Syracuse, New York, where she is a SUNY Distinguished Teaching Professor of Environmental Biology, and the founder and director of the Center for Native Peoples and the Environment.

ROOTED BEINGS

This Book is a Plant expands the ideas and ethos of Rooted Beings, an exhibition at Wellcome Collection, London (24 February–29 August 2022) curated by Bárbara Rodríguez Muñoz.

Rooted Beings explores our symbiotic relationship with the vegetal world through the work of artists Patricia Domínguez; Gözde Ilkin; Ingela Ihrman; Joseca; Eduardo Navarro; Resolve and Sop; alongside collection displays from Wellcome Collection's rich botanical collections and Kew Garden's herbarium specimens. The exhibition's three lines of enquiry – 'symbiosis', 'colonial violence & indigenous knowledge' and 'wilding' – aim to slowly decompose the artificial yet pervasive wall between us and nature that is devastating our ecosystems, our liveliness and our health. We invite you to embark on a meditative reflection on the vegetal world and what we can learn from it, to attain vegetal enlightenment.

Rooted Beings is a collaboration with La Casa Encendida, Madrid.

WELLCOME COLLECTION publishes thought-provoking books exploring health and human experience, in partnership with leading independent publisher Profile Books.

WELLCOME COLLECTION is a free museum and library that aims to challenge how we think and feel about health by connecting science, medicine, life and art, through exhibitions, collections, live programming, and more. It is part of Wellcome, a global charitable foundation that supports science to solve urgent health challenges, with a focus on mental health, infectious diseases and climate.

wellcomecollection.org